CAFE' LEBUG

sam humphrey

Inspirit Alliance

Café leBug

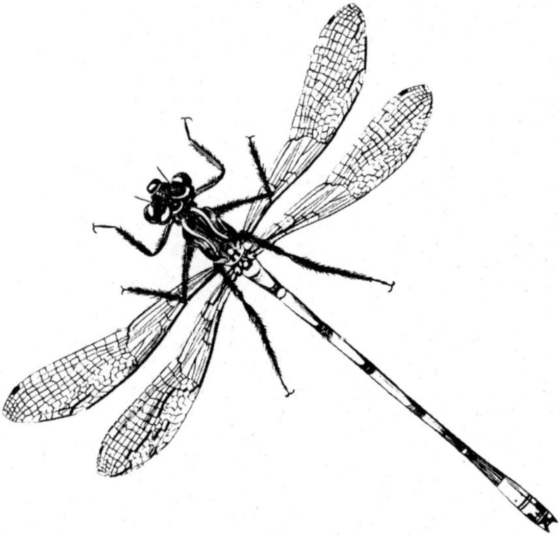

by sam humphrey

First Edition

Paperback ISBN-13: 978-1-959239-07-9
Hardback ISBN-13: 978-1-959239-09-3
Library of Congress Control Number: 2023904738

Cover design by: sam humphrey
Interior design by: sam humphrey

Printed in the United States of America

This book is dedicated to

Sawyer

and all the other people

Young and Old

Who find the life on this

planet interesting and

worth a second look.

You can watch a Bug crawling around
for minutes or hours at a time,
just sit and marvel at its complexity,
the utter Bugness of it

Contents

FOREWARD

Cafè leBug was originally a feature page in the little-known monthly magazine called **The Wolf Creek Journal** designed, edited, and published by Cindy Humphrey from 2000 to 2007. Cindy did all her work on the journal from her internet cafè The Wolf Creek Cyberstop in Wolf Creek, Montana. In 2002 sam humphrey started helping her out with the production of the little magazine. In 2005 sam started writing a page called "Cafè leBug" that featured one bug per issue.

Sam invited the readers to request bugs to be featured. He would then research the insect or creepy crawly and do a little write-up on the creature. He would find a drawing or two of the subject and include it in the article.

This book is the compilation of those articles with some new bugs included. Although the target audience is young adults, it is hoped that people of all ages will enjoy this book. The drawings in this book are included in a larger format so that the readers may enjoy coloring the bugs and bringing them to life. It is hoped that children, teens, and adults will all enjoy all aspects of this book. Share this book with your family and your friends. I ask you to read it to your children, and let them color the drawings. I have included several blank "What am I?" identification charts for you to fill out with bugs that you find interesting and want to learn more about. They don't have to be filled out perfectly - it is a learning experience. I have also included a small glossary at the back of the book. Most of the words typed in **bold** will be found in the glossary.

It is my hope that people will use this book to look around their world a little closer and get to know some of the smaller forms of life that we share this planet with.

Thank you, sam humphrey

WHAT IS A BUG?

What is a BUG? Is it a car? Sometimes. Is it a virus or a bacterium that makes you sick? Is it an action? You know, like "That really bugs me!" or "Man, don't BUG me!" Or is it something else? Well, in this book we are looking at the BUG as an insect, spider, worm, and other things like that.

When we say 'Bug' we are usually looking at some sort of creepy-crawly like arthropods with at least 6 legs, such as insects, spiders, and centipedes. Sometimes we refer to worms as "bugs" also, but they really aren't a bug. In fact, a true "BUG" is an Insect in the group Hemiptera – it must have piercing mouthparts and 6 legs. All Bugs are Insects, but not all Insects are Bugs. Spiders are not Bugs. Spiders are Arachnids and have 8 legs. Centipedes, Millipedes, and Sowbugs are not Bugs. They're not even Insects. Like Spiders, they are Arthropods. Unlike Spiders, they are not Arachnids – they fall into another Class of creature. Worms are all together in a whole different Phylum.

Sorry to be so scientific – you don't have to be this scientific to look at and enjoy Bugs, but it can be interesting. All living creatures on this planet are divided into groups. Those groups are made up of smaller groups. Those smaller groups are made up of even smaller groups. We call this process Taxonomic Identification or simply just Taxonomy.

The order goes like this: Kingdom, Phylum, Class, Order, Family, Genus, and finally *Species*. If we look at us – Human Beings – we are classified like this: Kingdom-Animal, Phylum-Chordata, Class-Mammal, Order-Primate, Family-Hominid, Genus-Homo, *Species-Sapiens*. We call ourselves Homo *sapiens*. Each one of those groups means something and narrows down to a specific life form with certain traits. Biologists – and there are all kinds; Zoologists, Entomologists, Botanists, and many more that study specific forms of life – have even more groups than what I am presenting here. There is no need to delve too deep into the science of it. I just want to introduce you to the process. If nothing else, it's good for trivia!

For each of the "Bugs" featured in this book I have an Identification chart that includes the Taxonomy of the creature described. This is just to show you how it is written down. When you study biology you will learn the process. It involves lots of questions like: Is it alive? Does it have a backbone? Is it warm-blooded like us, or is it cold like a fish? Does it have a tail? And on and on.

When you look at bugs maybe you could have questions like this in the back of your mind. Questions we ask about bugs are: How many legs does it have? Does it have a hard body or a soft body? Does it have wings? How many body sections does it have? Answers to those questions help us figure out what kind of bug we are actually looking at.

I have included a couple charts – pictures – that show the basic lay out of bugs parts so you will have an idea what I am talking about as I describe the bugs included in this book. I hope that helps.

Insect parts

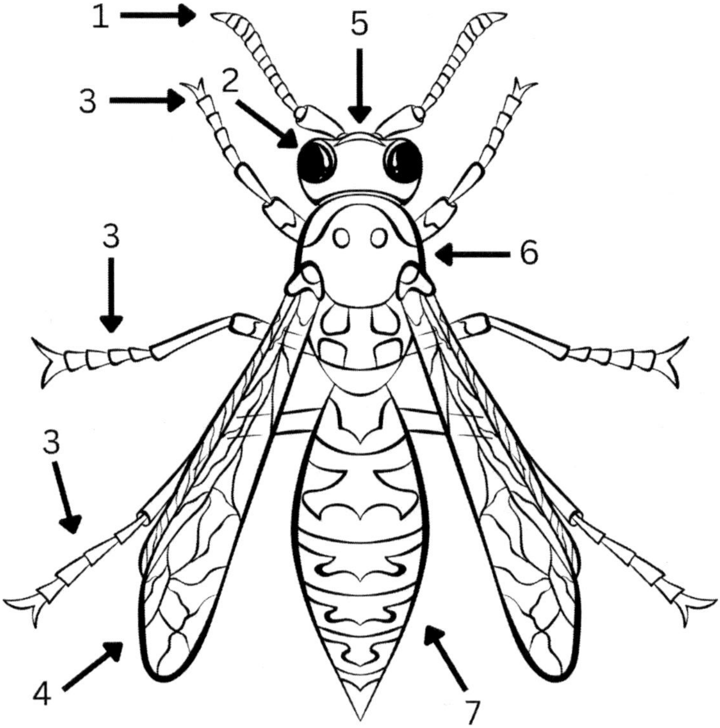

1. Antenna
2. Eye
3. Leg
4. Wing
5. Head
6. Thorax
7. Abdomen

Spider parts

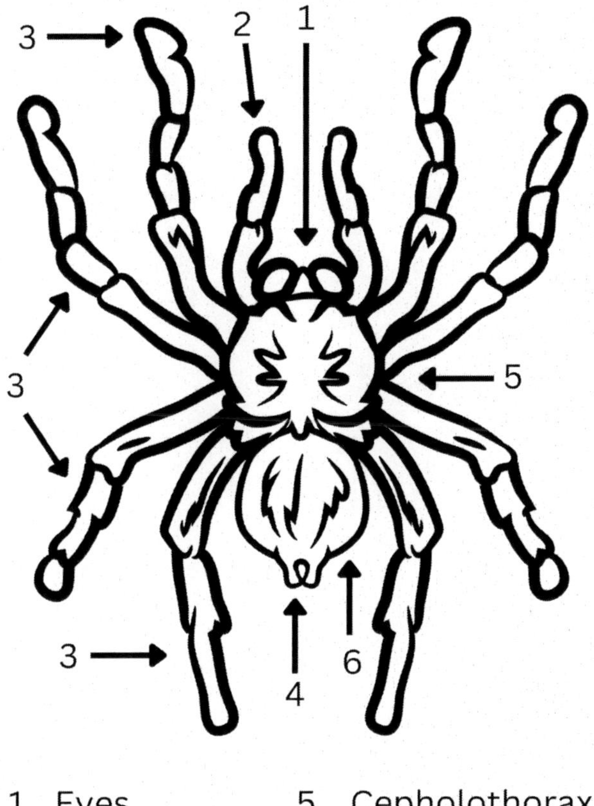

1. Eyes
2. Palps
3. Legs
4. Spinnerets
5. Cepholothorax
   Head & Thorax combined
6. Abdomen

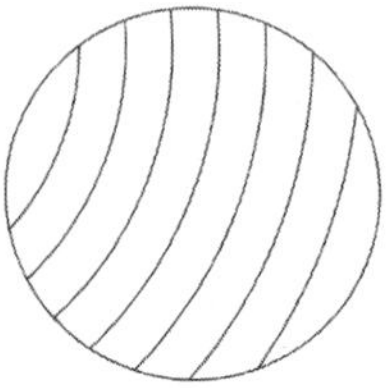

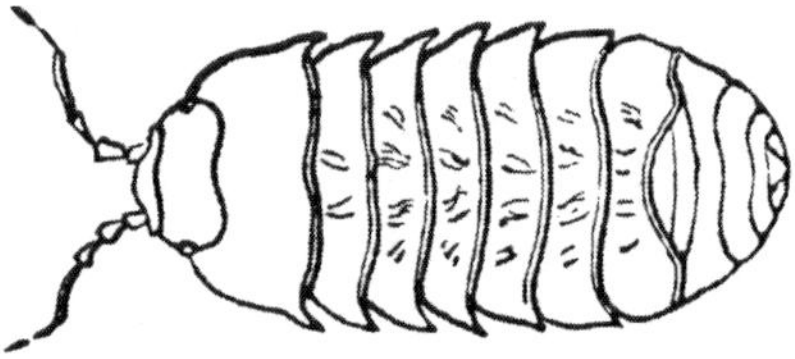

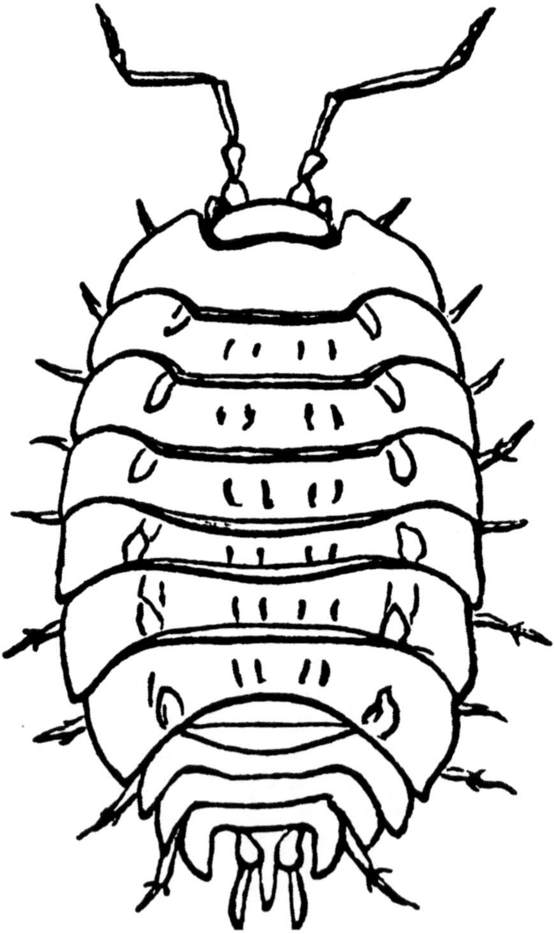

Pill Bug

For the first entreè in this book we will serve up the Pill Bug. It is known by several common names including Roly-Poly, Armadillo Bug, and Woodlice. It is not an insect or a "true bug", but rather an **Isopod**, meaning "uniform foot" or "all the feet are the same." There are two types of **Isopod**; **Aquatic** and **Terrestrial**. Pill bugs are **terrestrial**, meaning that they live on land. Sow bugs and gribbles are also in this group. These are in the class **Crustacea**, the same as crabs and shrimp.

Adult pill bugs average 3/4 of an inch in length. They are usually a dull grey to earthy brown in color. Their body is made up of the head, **thorax** and a five-segmented **abdomen**. Their body is usually fused together making it very hard to see where the head ends or the **abdomen** starts. They walk around on seven pairs of legs and they have a pair of **vestigial** antennae protruding from the front of their head. They also have two **compound** eyes on the top front of their head.

Pill bugs are monogamous and live in burrows in single family units. The male guards the entrance to the burrow and all the members of the family work together to keep the burrow clean. The female has a marsupium - a brood pouch - that she carries the eggs in until they hatch. Both the adult male and female gather food for the entire family group. A typical pair can have two broods a season. After the offspring mature, they move out into their own burrows; usually within three hundred feet of the parent's burrow. Pill bugs can live up to five years.

Pill Bugs are **omnivorous**, eating both vegetable and animal matter. Their diet normally consists of fungi, dead plant matter and animal waste. However, if their normal diet is scarce, they have been known to eat live plant matter.

Rolling into a ball or "conglobulating" is a defense mechanism for these creatures. They also hide from the light, are most active at night, and depend on their coloration to help camouflage them. Pill bugs typically live in moist or humid environments such as basements or under rocks, logs and other debris. They have a gas-permeable shell and can lose moisture through their shell, so they are susceptible to desiccation - drying out. They can't live in water though as they are air breathers and would drown.

What am I? Pill Bug

What is my Body Type?

Hard ☐ Segmented ☒ How many Segments?

Soft ☒ Worm like ☐ 2 ☐ 3 ☐ More X

Do I have Legs? Yes ☒ How many? 14 No ☐

Do I have Wings? Yes ☐ How many? ______ No ☒

Are my Wings under ***Elytra***? Yes ☐ No ☒

Do I have Stripes? Yes ☐ No ☒ How Many? ____

Do I have Spots? Yes ☐ No ☒ How Many? ____

What Color am I? grey My Spots or Stripes? ______

Kingdom Animalia

Phylum Arthropoda

Class Malacostraca

Order Isopoda

Family Calliphoridae

Genus Armadillidiidae

Species A. vulgare

Draw Me

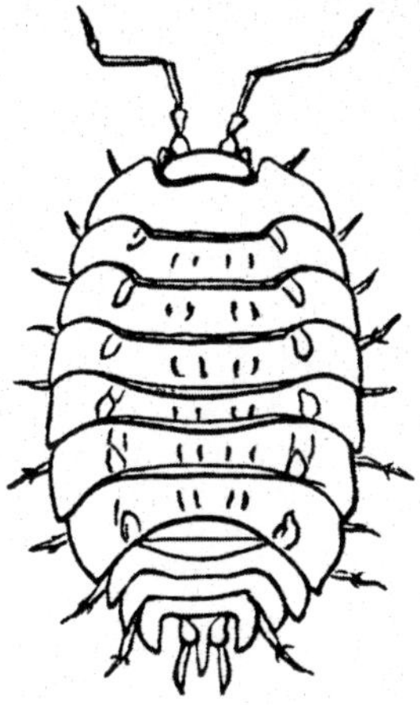

Ladybug

"Ladybug, ladybug fly away home. Your house is on fire and your children will burn!" Or so we said when I was a kid. Here we look at the ever-popular Ladybug. The most common and well-known ladybug is the Convergent Lady Beetle or Hippodamia *convergens* to the scientists in our group. Ladybugs are members of the family Coccinellidae in the order Coleoptera. These are considered beneficial insects because they predate, or eat, other insects that we consider harmful.

Both adult and juvenile ladybugs eat bugs, mostly aphids. They are known to eat cotton, pea, melon, cabbage, potato, green peach, and corn leaf aphids, as well as asparagus beetle eggs and larvae, and potato psyllids. If any of these insects are not to be found the ladybug will eat other small insect larvae, eggs and mites. They may also feed on the honeydew that is secreted by aphids and other sucking insects.

Ladybugs come out in the spring to mate and begin laying eggs. A typical female can lay over a thousand eggs during the summer. She lays her eggs in small clusters near aphid, or other prey, populations. The egg clusters can be found in protected sites on leaves and stems. The eggs are small and spindle-shaped.

The **larvae** of the ladybug are dark and "alligator-like" with three pairs of legs. The **larvae** start out small, less than a sixteenth of an inch, and will grow to over a quarter-inch long in 10 to 30 days. Then they go into a **pupal** stage that can last from 3 to 12 days, depending on the temperature.

The adult ladybug is one-quarter to three-eighths of an inch long. Its head is black and they have white markings just behind the head. The back is covered with a red **fore-wing** that may have only one or two black spots or as many as thirteen spots.

Ladybugs may go through one or two generations in a summer. Adults may live as little as a few weeks to as much as several months, depending on food availability, temperature and other environmental factors. Typically, the ladybugs that reach maturity in the late summer will go into hibernation in the fall. The adults congregate in large groups in the mountain valleys far from food sources to hibernate. They typically hibernate for nine months, from September to May, then re-emerge the following spring and disperse searching for good sites to lay eggs and feed.

Ladybugs are found from southern Canada, through-out the continental U.S., central America and all the way down deep into south America. Their main habitats are crop fields, orchards and vegetable gardens. They can be found in greenhouses or anywhere there are populations of aphids.

Ladybugs don't bite people, or any other larger animals. They are not harmful to any of our crops, buildings, or livestock. They are a "beneficial bug" because they are a natural pest control.

What am I? Ladybug

What is my Body Type?

Hard [X] Segmented [X] How many Segments?

Soft [ ] Worm like [ ] 2 [ ] 3 [X] More ___

Do I have Legs? Yes [X] How many? 6 No [ ]

Do I have Wings? Yes [X] How many? 2 No [ ]

Are my Wings under ***Elytra***? Yes [X] No [ ]

Do I have Stripes? Yes [ ] No [X] How Many? ___

Do I have Spots? Yes [X] No [ ] How Many? 9

What Color am I? red My Spots or Stripes? black

Kingdom Animalia

Phylum Arthropoda

Class Insecta

Order Coleoptera

Family Coccinellidea

Genus Coccinella

Species novemnotata

Draw Me

What am I? ____________

What is my Body Type?

Hard ☐ Segmented ☐ How many Segments?

Soft ☐ Worm like ☐ 2 ☐ 3 ☐ More ___

Do I have Legs? Yes ☐ How many? ______ No ☐

Do I have Wings? Yes ☐ How many? ______ No ☐

Are my Wings under ***Elytra***? Yes ☐ No ☐

Do I have Stripes? Yes ☐ No ☐ How Many? ____

Do I have Spots? Yes ☐ No ☐ How Many? ____

What Color am I? ______ My Spots or Stripes? ______

Kingdom __________

Phylum __________

Class __________

Order __________

Family __________

Genus __________

Species __________

Draw Me

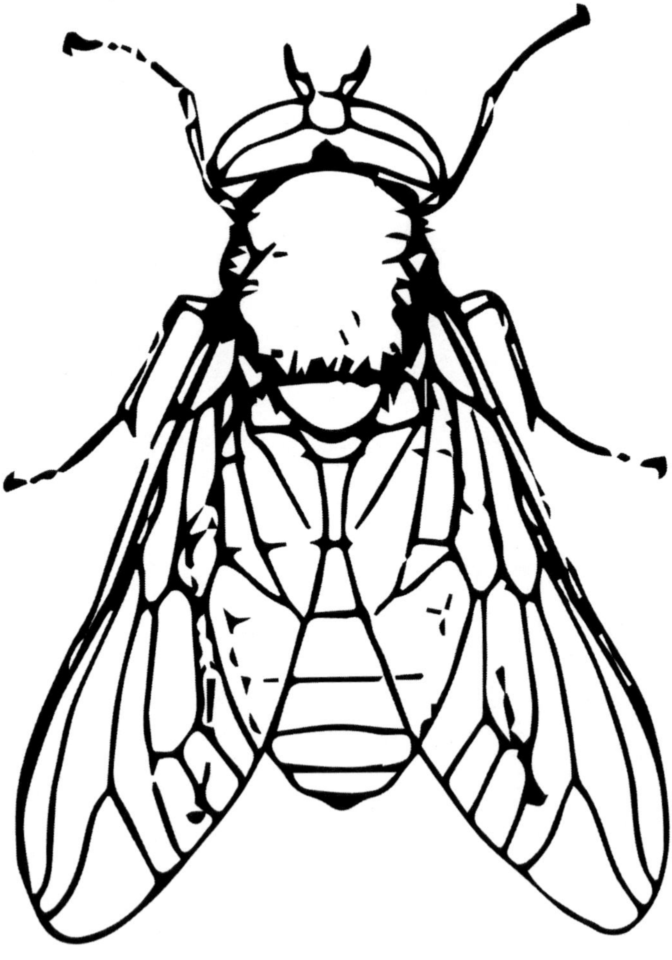

Cluster Fly

Here is an insect that lives up to the misnomer "BUG" and does it well. In the winter months this insect (or trio of insects really) bugs the heck out of me and probably you too! It is commonly called the Cluster Fly or sometimes the Attic Fly - some people may even call it a house-fly, but it isn't. It is one of 3 flies (or maybe even all 3 if your unlucky enough!), namely, the Cluster Fly (Pollenia *rudis*), or the Face Fly (Musca *autumnalis*), or the Blow Fly-2 types-(Phormia *sp*; Lucilia *sp*.). These flies all overwinter inside buildings if they can. They normally enter buildings in the early fall and come out to buzz around on warm, sunny days or anytime the building is a little warmer. They tend to gather at windows and often in rooms that are not used often. None of these flies eat or reproduce in the house, nor do they pose a health risk.

Cluster Flies are slightly larger than house-flies and move sluggishly. They have short golden hairs on the **thorax**. This fly lays its eggs on earthworms in the summer. The larvae develop as **parasites** of the worm. In late summer and early fall the adults emerge and seek a place to overwinter.

Face Flies are those super-annoying insects that irritate cattle and horses by swarming around the eyes. The **larvae** of this beast develop in the fresh manure of stock animals. The adults emerge in the summer and feed on the secretions of mucous from the eyes and noses of horses and cattle. This fly looks just like the common house-fly and, in fact, only an expert can tell the difference.

There are 2 types of Blow Fly that fall into this group of annoying winter in-house insects. They are blue or green bottle flies. They can be identified by their shiny metallic bodies. They can be fairly large and quite robust also. The adults lay their eggs in manure and **carrion** and usually show up in small numbers, but they can be everywhere! These are strong fliers and are attracted to light. These are the ones that buzz, buzz, buzz around the lamps, lights and windows like they're on a mission. They always seem to know when you are after them with a swatter too! Oooh, they really bug me!

These flies enter the house or buildings through cracks or spaces in the walls, eaves, around windows and doors and around the basement. They hibernate in the voids in walls and other dark spaces.

You can diminish their numbers with sprays of 0.5% diazinon or 3% malathion applied to inside surfaces where they tend to congregate. You can also use these sprays on the outsides of the buildings to decrease the numbers of those flies seeking places to spend the winter.

What am I? Cluster Fly

What is my Body Type?

Hard ☐ Segmented ☒ How many Segments?

Soft ☒ Worm like ☐ 2 ☐ 3 ☒ More ____

Do I have Legs? Yes ☒ How many? 6 No ☐

Do I have Wings? Yes ☒ How many? 2 No ☐

Are my Wings under ***Elytra***? Yes ☐ No ☒

Do I have Stripes? Yes ☒ No ☐ How Many? 4

Do I have Spots? Yes ☐ No ☒ How Many? ____

What Color am I? tan My Spots or Stripes? black

Kingdom Animalia

Phylum Arthropoda

Class Insecta

Order Diptera

Family Calliphoridae

Genus Pollenia

Species P. rudis

Draw Me

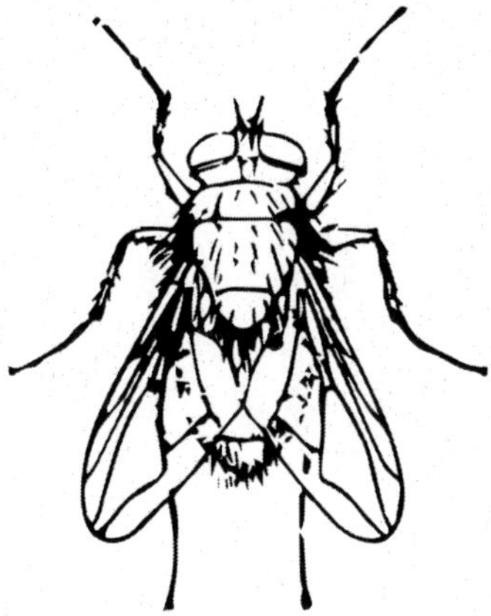

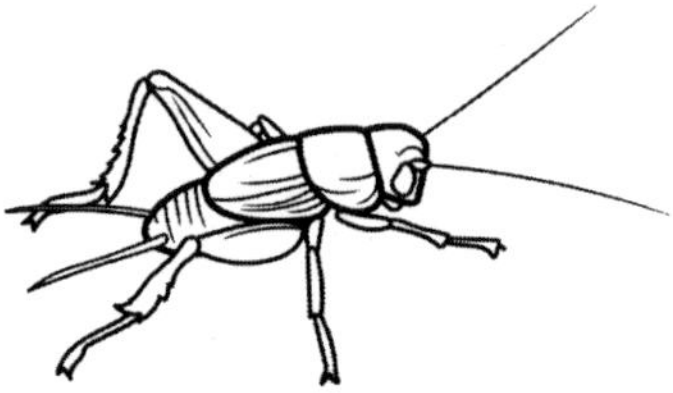

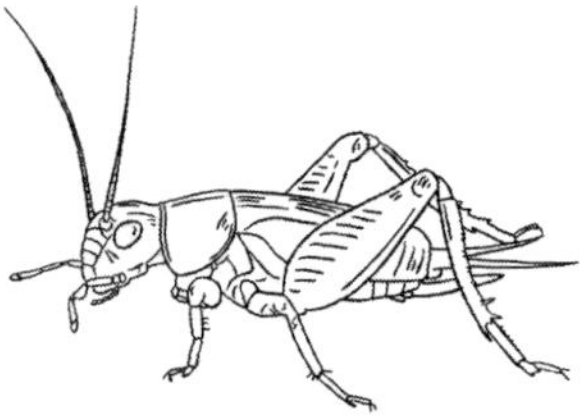

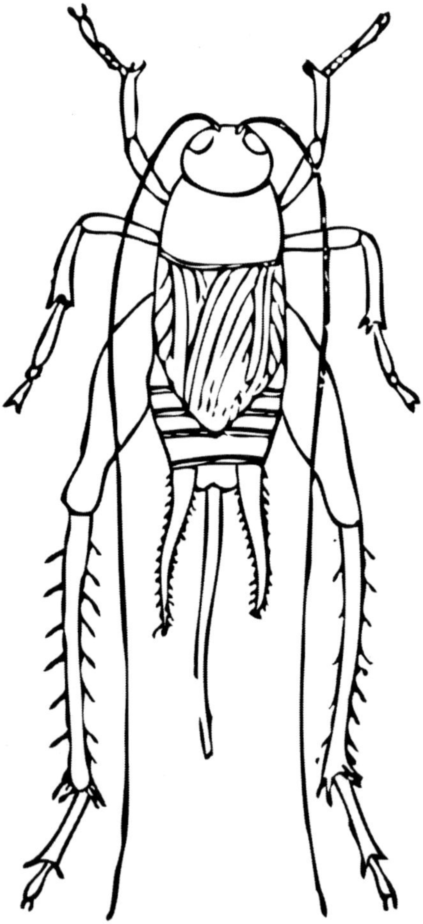

Cricket

For your pleasure we serve up the Cricket. Of the many types of crickets that exist there are two most commonly encountered in north America and Europe. These are the House Cricket (Acheta *domesticus*) and the Field Cricket (Gryllus *bimaculatus*).

House crickets are golden tan to pale brown in color with a black pattern on the head and **thorax** (usually three black stripes). The adults are about one inch long. Field crickets are variable in size ranging from a half-inch to over 1 inch. Most field crickets are black, but can be various shades of brown or tan.

Crickets are closely related to grasshoppers and locusts, and like them, have large hind legs which are used for jumping. Their legs have short, sharp spines on them. The **tympanum**, or ear, is located on the **tibia** of the hind legs. Only the males stridulate, or chirp, by rubbing the teeth on the sharp edge of one wing against a thick and rough scraper on the opposite wing. They have two pair of wings which are held flat over their backs, but only the back pair are used for flying. Both male and female crickets have slender antennae that are longer than their bodies and both also have long antenna-like feelers at the rear-end of their body. Females also have a long slender egg-laying guide in between the two rear feelers called an **ovipositor** that allows them to lay eggs in the ground.

Crickets take two to three months to complete their life cycle, depending on temperature. The warmer their climate, the faster they mature and the more eggs they lay. Eggs are deposited in whatever damp substrate is available. Each female will lay between 50 to 100 eggs that hatch in two to three weeks. Newly hatched **nymphs** are the same size as the eggs, and blend in with their surroundings. The young crickets look like adults, except that they are smaller, have no wings and the genital organs have not developed. The **ovipositor** does not develop until the last few moltings and gets longer with each successive **molt**. It takes these tiny crickets eight to twelve weeks to reach full maturity. Adult crickets live two to three months.

All crickets have chewing mouthparts. They feed on decaying leaves, living plants, and other insects. Indoors, crickets can feed on food particles, cloth, leather and fur. They mostly feed on items that have food, beverages or perspiration on them. They feed on the surface of fabrics, such as cottons, silks or woolens, leaving roughened areas caused by their picking and pulling to loosen the fibers.

Lots of info is available out there on crickets, they have been raised by humans for several thousand years. Be careful when searching the internet for info on the insect as there is lots of info on the English game of cricket too! Have fun investigating!

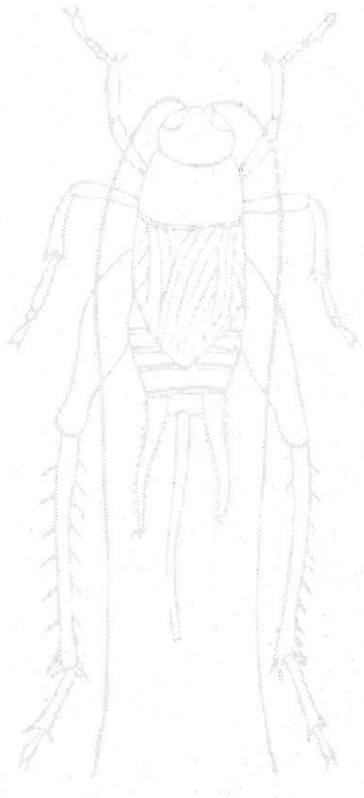

What am I? Cricket

What is my Body Type?

Hard [X] Segmented [X] How many Segments?

Soft [ ] Worm like [ ] 2 [ ] 3 [X] More ___

Do I have Legs? Yes [X] How many? 6 No [ ]

Do I have Wings? Yes [X] How many? 4 No [ ]

Are my Wings under ***Elytra***? Yes [X] No [ ]

Do I have Stripes? Yes [X] No [ ] How Many? 3

Do I have Spots? Yes [ ] No [X] How Many? ___

What Color am I? tan My Spots or Stripes? black

Kingdom Animalia

Phylum Arthropoda

Class Insecta

Order Orthoptera

Family Gryllidae

Genus Acheta

Species domesticus

Draw Me

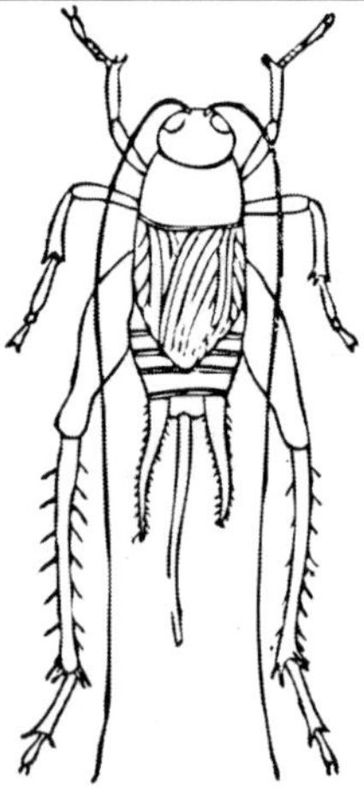

What am I? ______________

What is my Body Type?

Hard ☐ Segmented ☐ How many Segments?

Soft ☐ Worm like ☐ 2 ☐ 3 ☐ More ___

Do I have Legs? Yes ☐ How many? ______ No ☐

Do I have Wings? Yes ☐ How many? ______ No ☐

Are my Wings under ***Elytra***? Yes ☐ No ☐

Do I have Stripes? Yes ☐ No ☐ How Many? ____

Do I have Spots? Yes ☐ No ☐ How Many? ____

What Color am I? ______ My Spots or Stripes? ______

Kingdom __________

Phylum __________

Class __________

Order __________

Family __________

Genus __________

Species __________

Draw Me

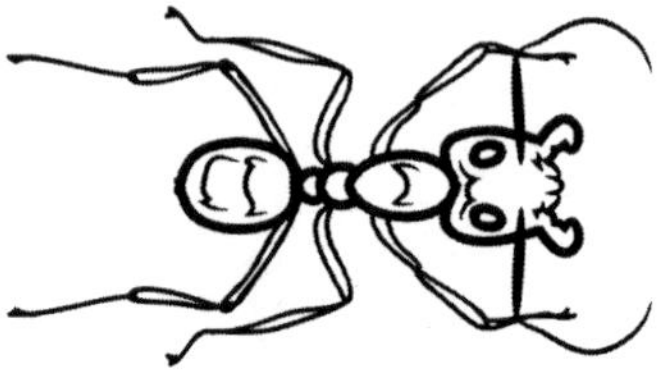

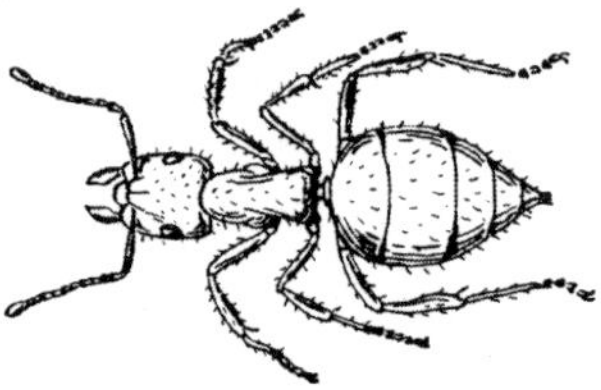

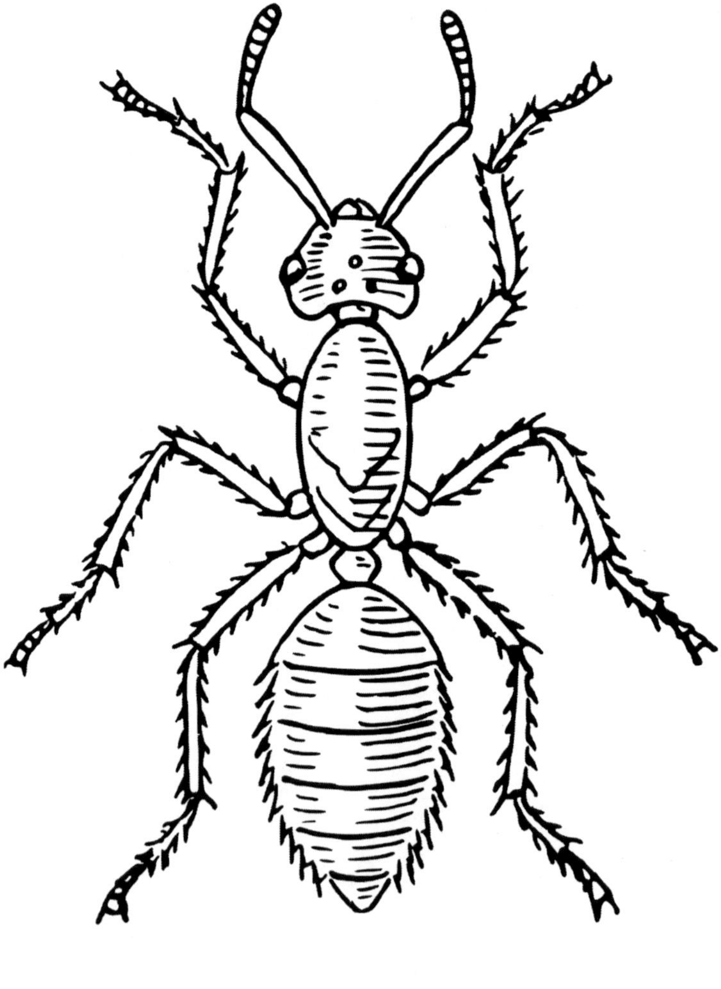

CARPENTER ANT

Crawling around on this page is an insect that can be found almost everywhere, the Carpenter Ant. Several types of this ant may be seen in the spring. The worker is a large, black female ant (1/4 to 5/8 of an inch) that roams around indoors and outdoors looking for food. The young queen is a winged female ant that is in search of a site to start a new colony. The drone is a winged male ant that looks for new queens to conduct mating flights with. An established queen may grow to 3/4 of an inch, but will not leave the nest.

Carpenter ants are active in the warmer months. They are usually dormant in the winter months, unless they have a nest in a warm building. So if you have ants crawling around inside your house during the cold winter months, it is likely that they have nested somewhere inside your place. They do not actually eat would so the damage they cause is not as immediately severe as damage from termites. They chew out chambers or galleries and leave the sawdust outside the nest. Colonies can exist for many years, decades even, so the damage a single colony can cause increases with time as the actual size of the colony grows.

A colony is started by a new queen from an established colony. The new queen performs a mating flight with a drone then finds a suitable place to start her new colony, usually in deteriorating, moist wood such as rotting logs, stumps, dead trees and live trees. Common places for nests around houses are window sills, roofs, porch pillars, or any wood that is in contact with the ground. As colonies grow they may move into sound wood that is near the existing colony. The queen lays the eggs of her first brood and raises them on **salivary** secretions. This brood matures into workers who start foraging for food for the next broods. When a good food source is found the colony starts to grow rapidly. New workers begin chewing out the galleries and foraging for food for the growing colony. A colony does not start producing new queens or drones until its own population is around two thousand individuals. This can take anywhere from three to six years. At this point the colony will keep producing drones and queens.

As stated before, carpenter ants do not eat wood, instead they feed on the honeydew from aphids and other insects. They also feed on plant juices, and will also eat food scraps found inside houses and picnic areas. They may also search for water inside houses.

Carpenter ants are in the order Hymenoptera, family Formicidae and genus Camponotus. There are hundreds of species of carpenter ant found in the United States. This can make proper identification a little difficult.

What am I? Carpenter Ant

What is my Body Type?

Hard [X] Segmented [X] How many Segments?

Soft [ ] Worm like [ ] 2 [ ] 3 [X] More ___

Do I have Legs? Yes [X] How many? 6 No [ ]

Do I have Wings? Yes [ ] How many? ___ No [X]

Are my Wings under ***Elytra***? Yes [ ] No [ ]

Do I have Stripes? Yes [ ] No [X] How Many? ___

Do I have Spots? Yes [ ] No [X] How Many? ___

What Color am I? black My Spots or Stripes? ___

Kingdom Animalia

Phylum Arthropoda

Class Insecta

Order Hymenoptera

Family Formicidae

Genus Camponotus

Species C. modoc

Draw Me

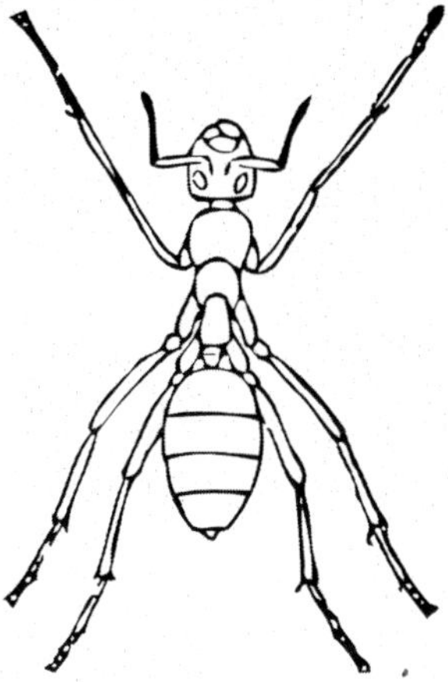

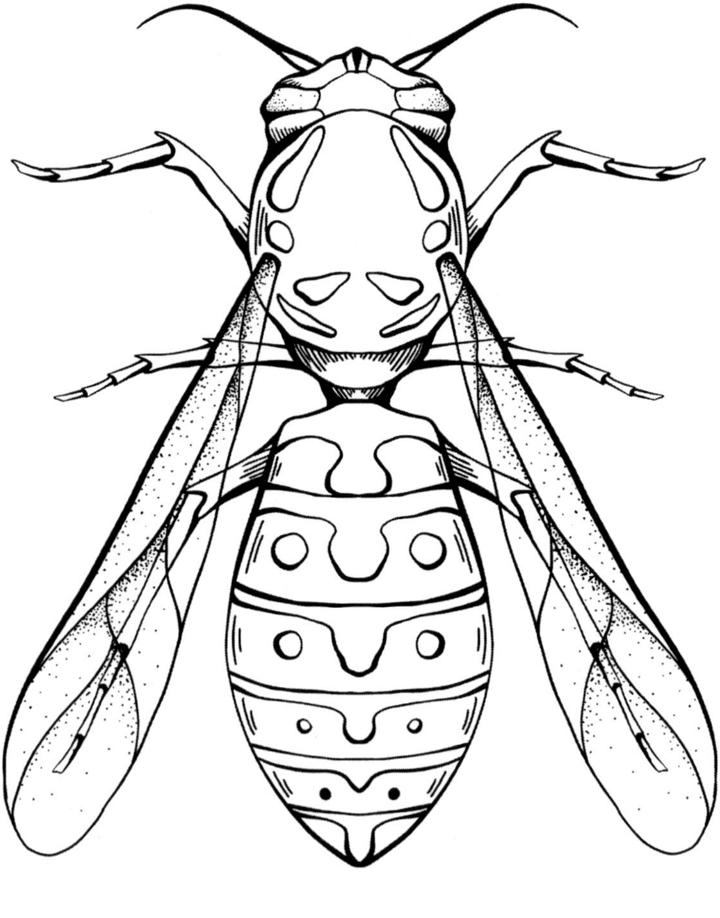

Bald Faced Hornet

Buzzing around this page is an insect that is all too familiar, but also commonly mistaken for other bugs. It is the Bald-Faced Hornet, entomologically known as Dolichovespula *maculata.* It is not really a hornet, but rather a wasp and is most commonly confused with its very close cousin the yellow jacket. However, the bald-faced hornet is an "aerial nester," meaning it builds its nest up off the ground - usually hanging in a tree, under a cliff overhang or under the eaves of buildings.

The Bald-Faced Hornet has a stout body. It is black with white markings on the head, **thorax** and rear half of the **abdomen**. The markings can be somewhat yellowish and not necessarily bright white. It has a thin neck and **pedicel** (or "waist"). The wings are translucent and smoky colored. the overall length of the wasp is from 5/8 to 3/4 of an inch, slightly larger than the Western Yellow Jacket.

The impregnated queens of this species come out of hibernation in the spring to locate a site for their nest. They build the first part of their nest and lay the first batch of eggs. This queen will do all the work of feeding and aggressively guarding her first brood, but after these mature into workers she will not leave the hive again and only lays eggs. The workers will take over the jobs of feeding, building and defending the colony. The colonies of the bald-faced hornet are small, usually having a population of 100 to 400 workers.

Bald-faced hornets, like other Hymenoptera, are haplodiploids, that is the males grow from unfertilized eggs and the females grow from fertilized eggs. The males are **haploid**, meaning they only have half a pair (a single set) of chromosomes. Females are **diploid**, meaning they have a full pair (or 2 sets) of chromosomes. The male's (drones) only purpose is to fertilize queens from other colonies.

The adults drink nectar and fruit juices. They may also eat other insects. The **larvae** are fed insects, mostly soft-bodied bugs such as caterpillars, that are pre-chewed by the adults.

When fall rolls around and the temperatures drop below freezing, the hive dies. All the workers die, the young, immature wasps die, the drones die and even the older queens die. Only the young fertilized females will live through the winter. These new queens will find a place to hibernate, usually in bark crevices or in the leafy and twiggy litter on the ground.

Bald-Faced Hornets are not aggressive, but they vigorously and viciously defend their hive. They will sting repeatedly (and not die!) and they have a particularly painful sting.

What am I? Bald Faced Hornet

What is my Body Type?

Hard [X] Segmented [X] How many Segments?

Soft [ ] Worm like [ ] 2 [ ] 3 [X] More ___

Do I have Legs? Yes [X] How many? 6 No [ ]

Do I have Wings? Yes [X] How many? 4 No [ ]

Are my Wings under ***Elytra***? Yes [ ] No [X]

Do I have Stripes? Yes [X] No [ ] How Many? 3

Do I have Spots? Yes [ ] No [X] How Many? ___

What Color am I? white My Spots or Stripes? black

Kingdom Animalia

Phylum Arthropoda

Class Insecta

Order Hymenoptera

Family Vespidae

Genus Dolichovespula

Species D. maculata

Draw Me

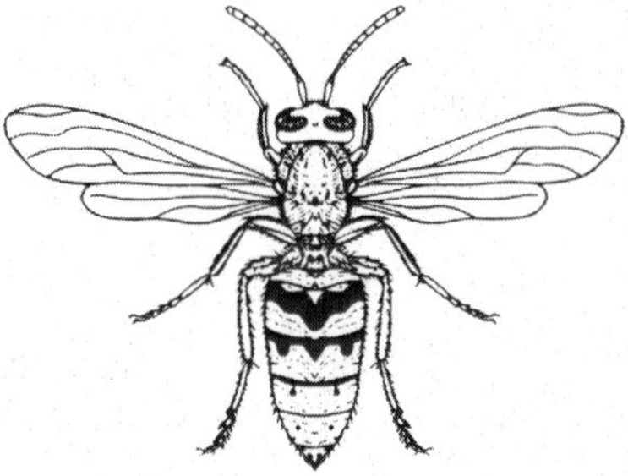

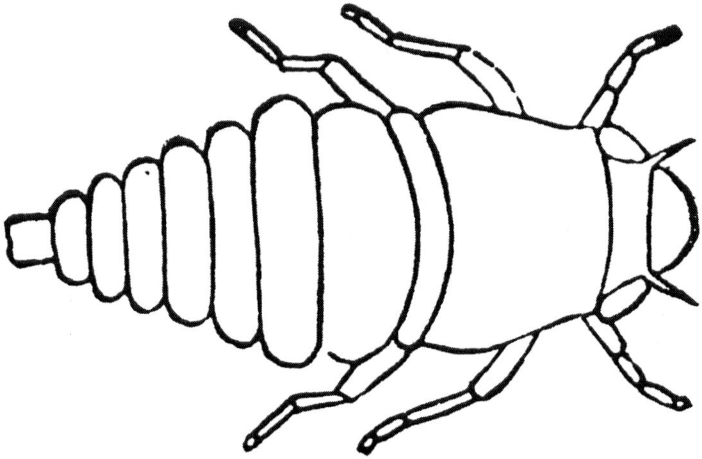

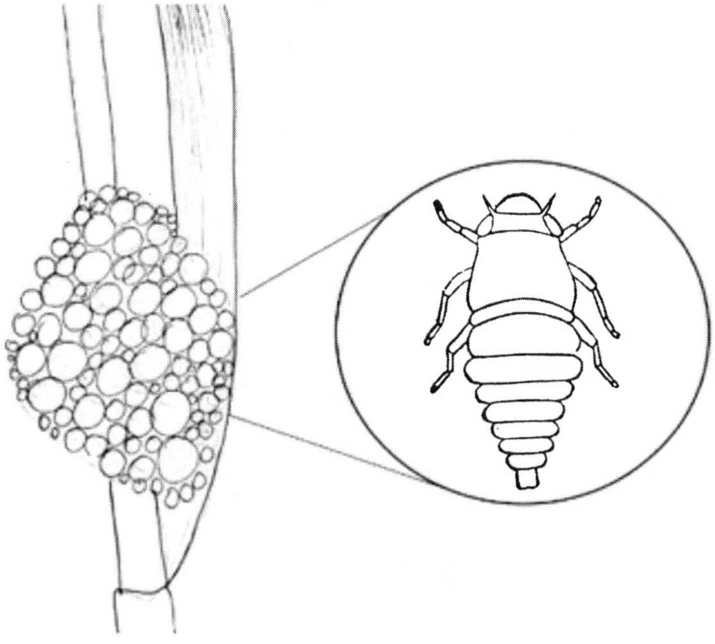

Spittlebug

The dish on these pages is served at the request of Cecelia McAfee. She saw some white foamy stuff on a plant, messed with it and discovered that there were bugs in it. Of course, she wanted to know all about it, so here is an introduction to the Spittlebug. There are over 23,000 species of spittle bug! Yes, that's right; **Twenty-Three Thousand** species! Holy Toledo! We have over 60 species of froghopper spittlebugs in North America – north of Mexico.

Spittlebugs are insects in the Order Homoptera (same wing), Family Cercopidae (froghoppers and spittlebugs). The common name, Spittlebug, comes from the bug's behaviour of making a foamy mass of bubbly liquid on plant stems that looks like spit. Ptui.

The "spittle" serves several purposes. It conceals the immature bug, or **nymph**, from predators, protects it from **parasites**, insulates it from the cold and keeps the **nymph** from drying out. The **nymph** penetrates plant stems with its sharp beak to suck the juices out. It digests these juices as food, then mixes the liquid with other chemicals, turning it into a soapy liquid that is excreted out its rear-end. This liquid trickles out until a sticky pool is formed, then the bug pumps air into the pool through an air tube protruding from its **abdomen**. It pumps its body like a bellows until it has completely covered itself in a mass of bubbles. This process takes ten to twenty minutes.

The spittlebug starts life as an egg laid by an adult female in the crook of plant leaves in the fall. In the spring the nymph hatches and makes a mass of protective bubbles. The nymph can move from plant to plant, but keeps itself covered with the spittle. The **nymph** goes through five **instars** before maturing into an adult. An **instar** is a stage of life in between **molts**. Molting is what creatures with hard bodies do as they grow. They **molt** – shed their hard shell – so their soft body can grow, after which they grow a new **exoskeleton** (outside skeleton). Wings emerge at the fourth instar stage, but are not fully developed or used until the adult stage. It takes anywhere from thirty to one hundred days for the insect to mature, depending on temperature. It normally takes one year for each generation, or life cycle to complete. As an adult the spittlebug has fully developed wings and strong rear legs for jumping, hence the common name "Froghopper." The adult is about 3/8ths of an inch long and pear-shaped. The adult varies in color from grey to green to yellow, even medium-brown with light spots. The nymph is pale and mostly colorless.

This insect can be quite destructive to crops, especially alfalfa and clover. It is considered a pest insect and is on the government's list of crop-damaging insects.

What am I? Spittlebug

What is my Body Type?

Hard [ ] Segmented [X] How many Segments?

Soft [X] Worm like [ ] 2 [ ] 3 [ ] More X

Do I have Legs? Yes [X] How many? 6 No [ ]

Do I have Wings? Yes [ ] How many? ______ No [X]

Are my Wings under ***Elytra***? Yes [ ] No [X]

Do I have Stripes? Yes [ ] No [X] How Many? ____

Do I have Spots? Yes [ ] No [X] How Many? ____

What Color am I? green My Spots or Stripes? ______

Kingdom Animalia

Phylum Arthropoda

Class Insecta

Order Hemiptera

Family Cercopidae

Genus Too many

Species Over 60

Draw Me

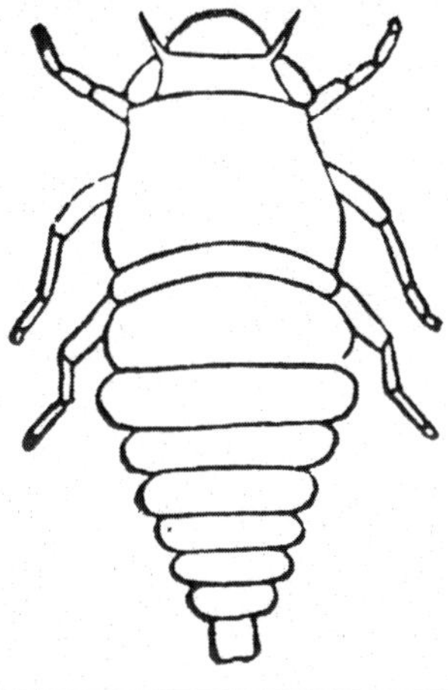

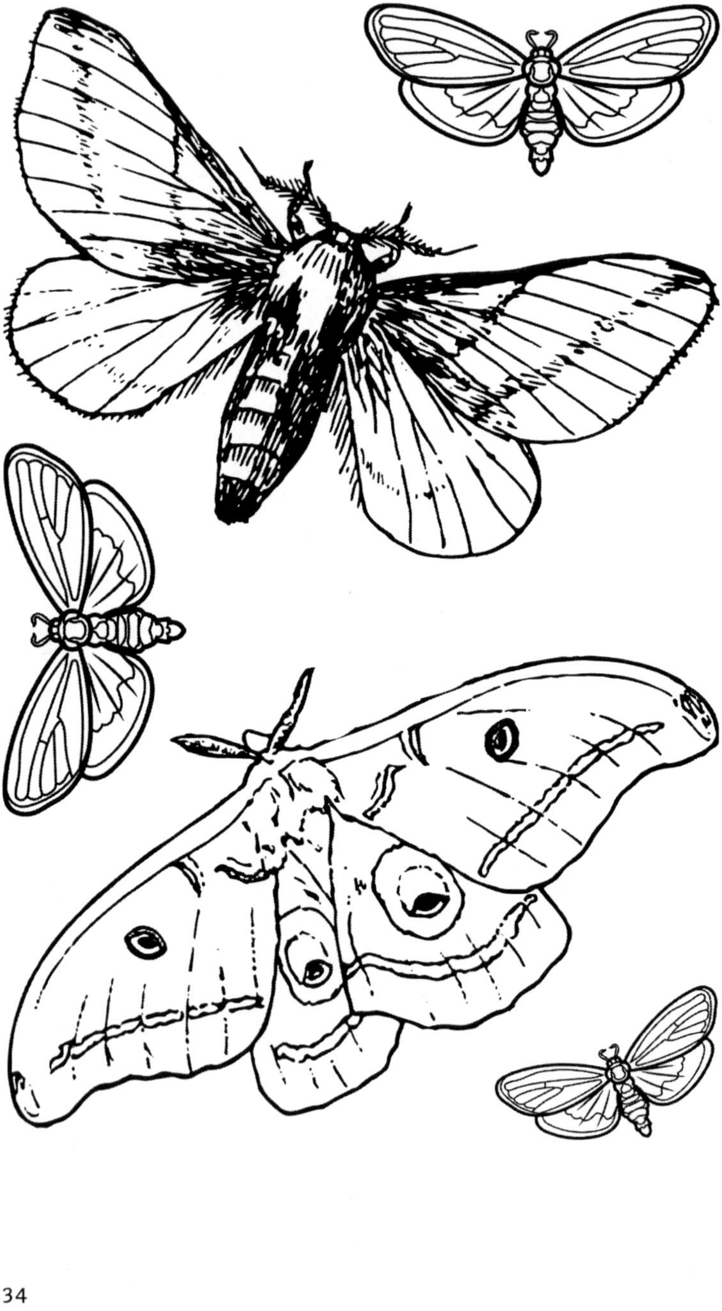

Miller Moth

The dish of choice on these pages is the Miller Moth. "Miller" is the common name for any species of moth found in abundance in and around homes. The name comes from the fine scales that rub off the wings of the moths reminiscent of the dusty flour that covered the clothes of grain millers. In our area (along with most of the rocky mountain states) the most common moth known as a Miller Moth is the Army Cutworm Moth, entomologically known as Euxoa *auxiliaris*. This cutworm is called an "army cutworm" because in large populations the caterpillars band together and move in large army-like groups across fields or highways.

These moths lay their eggs in the late summer or early fall at lower elevations, typically in areas of heavy plant growth, like the weedy spots of hay fields. The eggs hatch a few weeks after they are laid. The young caterpillars start feeding in preparation for hibernating over winter. In the spring they wake up and start feeding again. It is at this time that they do most of their damage to crops. The caterpillars become full-grown in mid-spring at which time they burrow into the ground to pupate. The **Pupal** stage lasts 2 to 3 weeks. The adult moths emerge from the ground and migrate. They can fly several hundred miles. They usually migrate from lower elevations to higher elevations. At the higher elevations the moths feed on nectar. They are not reproductively active nor do they lay eggs at this time. In the late summer and early fall the moths return to the lower elevations where they reproduce and lay eggs to start the cycle all over again.

The adult moths have wingspans of one-and-a-half to two inches. The wings are usually grey or light brown with wavy light and dark markings. The markings may vary greatly in pattern and coloration from moth to moth. The body of the adult moth is usually one to one-and-a-half inches long and about as big around as a pencil. The body color tends to blend in with the wing coloring.

The army cutworm caterpillars are medium to dark green in color with several lines of black spots along the body. There are black spots at the caterpillars front where the eyes and mouth are. The Pale Western Cutworm (Agrotis *orthogonia*) is a closely related species and is easily distinguished from the army cutworm in that the pale cutworm is light green in color and has no black spots. The front of the caterpillar is a lighter shade of green. Both of these caterpillars hide under dirt clods and debris during the day and feed from dusk to dawn and on cloudy days. These two caterpillars are responsible for quite a bit of damage to crops in the Rocky Mountain and Midwest states.

What am I? Miller Moth

What is my Body Type?

Hard ☐ Segmented ☒ How many Segments?

Soft ☒ Worm like ☐ 2 ☐ 3 ☒ More ____

Do I have Legs? Yes ☒ How many? 6 No ☐

Do I have Wings? Yes ☒ How many? 4 No ☐

Are my Wings under ***Elytra***? Yes ☐ No ☒

Do I have Stripes? Yes ☐ No ☒ How Many? ____

Do I have Spots? Yes ☒ No ☐ How Many? lots

What Color am I? brown My Spots or Stripes? black

Kingdom Animalia

Phylum Arthropoda

Class Insecta

Order Lepidoptera

Family Noctuidae

Genus Euxoa

Species E. auxiliaris

Draw Me

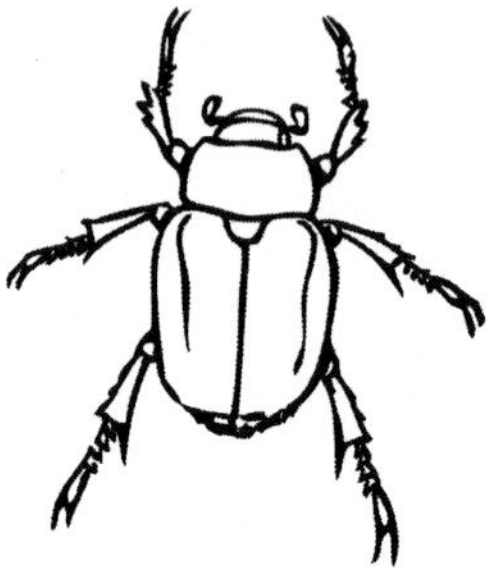

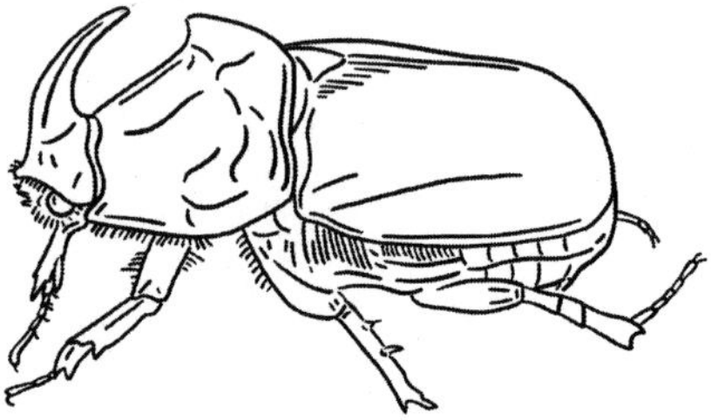

Beetle

Here we present the Beetle. No, not the VW Bug that so many of us drove back in the '60s and '70s. I mean insects of the order Coleoptera. I am not describing any one certain beetle, rather the whole group in general. Why would I describe a whole order of insects and not just a single species like I usually do here in the Cafe' LeBug? Well, it's like this: the other day I was riding my motorcycle back to Wolf Creek from Spokane and just before I pulled into Superior for gas, I was hit in the forehead by what felt like a rock. It knocked my head back and didn't feel very good. No, I wasn't wearing a helmet - Montana still allows some personal freedoms unlike many other states in the "land of the free." Anyway, Superior was less than a mile away so I kept going. When I got off my bike at the gas station, I saw an enormous (and I mean GIGANTIC) beetle on my seat. It must have landed there after hitting me in the head. It was dead. It was at least two-and-a-half inches long and three-quarters of an inch wide, dark brown with long antennae.

I thought this would make an interesting feature for our book. I started researching it on the web.

Problem is that the Order Coleoptera is so large, over 350,000 described species, that I couldn't find this particular beetle on the internet. That doesn't mean that it's not there, just means that there is so much information it would take me forever to wade through it. I did learn a lot of interesting things about beetles, though. I did see a lot of interesting, scary, even grotesque pictures, too.

One in five living species of all animals on this planet is a species of beetle. That means 20% of all known species now living are beetles. Fossil records show that beetles have been around since the Lower Permian Age or over 265 million years. Beetles range in size from less than a sixteenth of an inch to over eight inches in length. Most beetles are plant eaters, but some are predators (like ladybugs) that eat other insects or animals. There are species that eat only fungi and others that are **parasitic** (living off other life forms). The order Coleoptera has colonized every habitat known, from the rain forests to the deserts, from the tops of mountains above the tree line to the depths of lakes and oceans. There are, however, no known species inhabiting Antarctica. Beetles communicate by various means too. They may use sound (like click beetles), visual stimulation (like fireflies) or chemical (**pheromones**) communication.

The most distinctive feature of the beetle is its hardened **fore-wing**. In fact, Coleoptera means "sheathed wing." The body is made up of the head, **prothorax** and **pterothorax** combined with the **abdomen**. Genitalia is retracted into the abdomen. Adult antennae usually have eleven segments.

Human beings have been portraying beetles in art and making all kinds of jewelry that looks like beetles for thousands of years. The Scarab is a beetle that can be found all through ancient Egypt in art and jewelry. Egyptian hieroglyphs include the scarab as one of their symbols. Depictions of the scarab have even been found in the tombs of mummies!

I hope you have enjoyed this quick look at the beetle!

What am I? Beetle - June Bug

What is my Body Type?

Hard [X] Segmented [X] How many Segments?

Soft [ ] Worm like [ ] 2 [ ] 3 [X] More ____

Do I have Legs? Yes [X] How many? 6 No [ ]

Do I have Wings? Yes [X] How many? 4 No [ ]

Are my Wings under ***Elytra***? Yes [X] No [ ]

Do I have Stripes? Yes [X] No [ ] How Many? lots

Do I have Spots? Yes [ ] No [X] How Many? ____

What Color am I? green My Spots or Stripes? white

Kingdom Animalia

Phylum Arthropoda

Class Insecta

Order Coleoptera

Family Cetoniinae

Genus Polyphylla

Species decemlineata

Draw Me

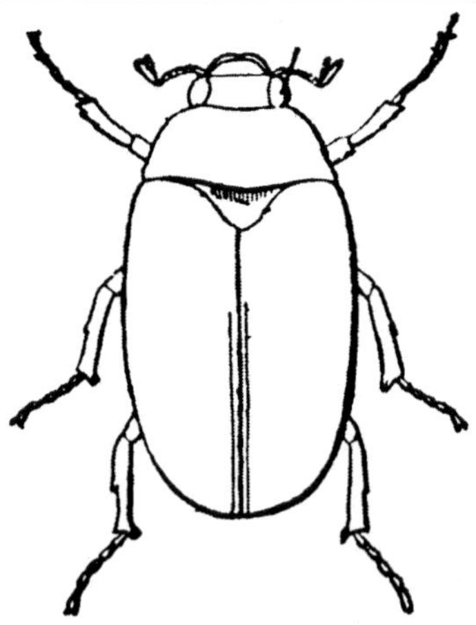

What am I? ____________

What is my Body Type?

Hard ☐ Segmented ☐ How many Segments?

Soft ☐ Worm like ☐ 2 ☐ 3 ☐ More ___

Do I have Legs? Yes ☐ How many? ______ No ☐

Do I have Wings? Yes ☐ How many? ______ No ☐

Are my Wings under ***Elytra***? Yes ☐ No ☐

Do I have Stripes? Yes ☐ No ☐ How Many? ____

Do I have Spots? Yes ☐ No ☐ How Many? ____

What Color am I? ______ My Spots or Stripes? ______

Kingdom __________

Phylum __________

Class __________

Order __________

Family __________

Genus __________

Species __________

Draw Me

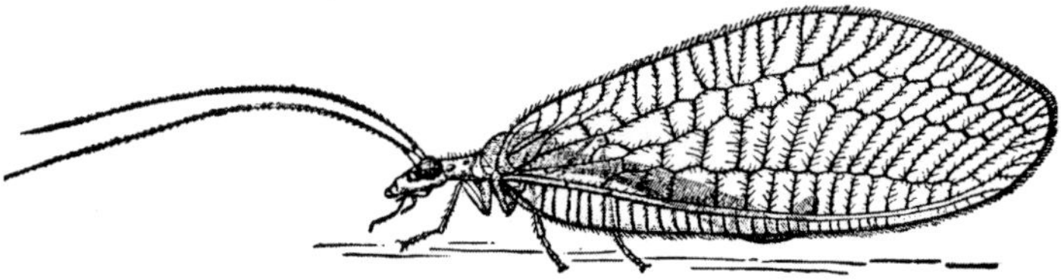

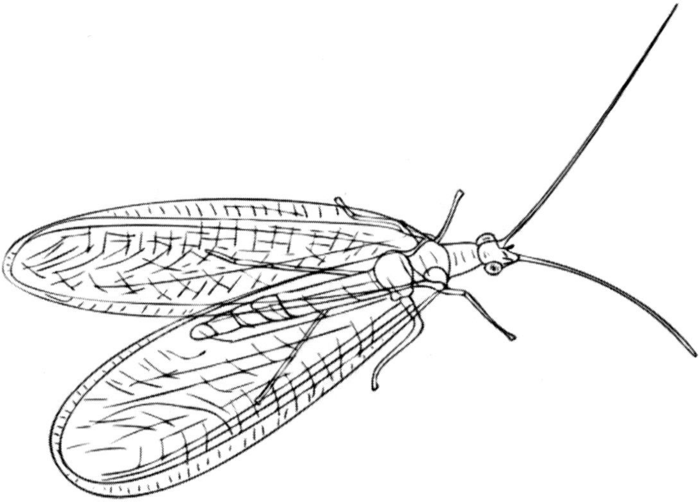

Green Lacewing

Here we are featuring the Green Lacewing (Chrysopa *carnea*), a delicate-looking bug that looks a little like a fairy when it is flying around. In reality, it is a **voracious** predator. The adult lacewing can show up inside your house in the winter. It, like many other bugs, overwinters inside buildings. It neither eats, nor reproduces during the winter months. In the spring and summer, the adult female lays eggs on leaves or twigs of plants near prey for the larvae. The female can lay several hundred eggs. She lays pale green, oval eggs singly. The eggs turn gray after a few days. The eggs are on stalks that hold them off the surface of the leaf or twig. This is possibly a strategy to prevent **cannibalism** from larvae that have already hatched.

The **larvae** emerge in 3 to 6 days and eat all bugs and eggs in sight. The very active **larvae** are brown and grey with well-developed legs and large pincers. The pincers are used to suck the body fluids from their prey. They prefer aphids (and are called "aphid lions"), but devour many soft-bodied insects, including spider mites, thrips, whiteflies, small caterpillars, and beetle **larvae**. They also consume the eggs of moths, leafhoppers and leafminers. They have been reported to eat as many as 60 eggs in an hour. In their larval stage they may consume up to 600 aphids. The larval stage includes 3 **instars** and lasts 2 to 3 weeks. When the third **instar** matures it spins itself into a round, silken cocoon that looks somewhat like a pea. 10 to 14 days later the adult emerges and starts the cycle anew. In summer conditions a complete life cycle can occur in as little as 4 weeks, so there can be several generations in a single year.

The adults have a delicate pale green body with long antennae and large, bright, golden-orange eyes. The wings are large and transparently pale green. The adults feed on the nectar of several types of plants. They are very active fliers, mostly at twilight and night to keep from being eaten by birds. They can also "hear" or sense bat sonar and stop flying to fall to the ground when bats are near. That way they won't get eaten by the bats.

Green lacewings also use body vibrations to communicate with each other, especially during courtship. Species very nearly the same can sometimes be told apart more easily by listening to their mating signals. For example, the southern European Lacewing looks almost identical to its northern relative, the Common Green Lacewing, but their courtship "songs" are very different. The individuals of one species will not react to the sounds made by a different species.

Look for those Lacewings remembering they are a "good bug" and don't kill them when you find them.

What am I? Green Lacewing

What is my Body Type?

Hard ☐ Segmented ☒ How many Segments?

Soft ☒ Worm like ☐ 2 ☐ 3 ☒ More ___

Do I have Legs? Yes ☒ How many? 6 No ☐

Do I have Wings? Yes ☒ How many? 4 No ☐

Are my Wings under ***Elytra***? Yes ☐ No ☒

Do I have Stripes? Yes ☐ No ☒ How Many? ___

Do I have Spots? Yes ☐ No ☒ How Many? ___

What Color am I? green My Spots or Stripes? ______

Kingdom Animalia

Phylum Arthropoda

Class Insecta

Order Neuroptera

Family Chrysopidae

Genus Chrysopa

Species C. carnea

Draw Me

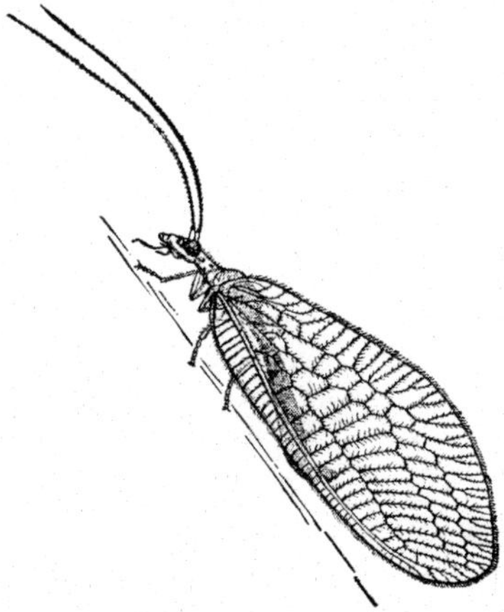

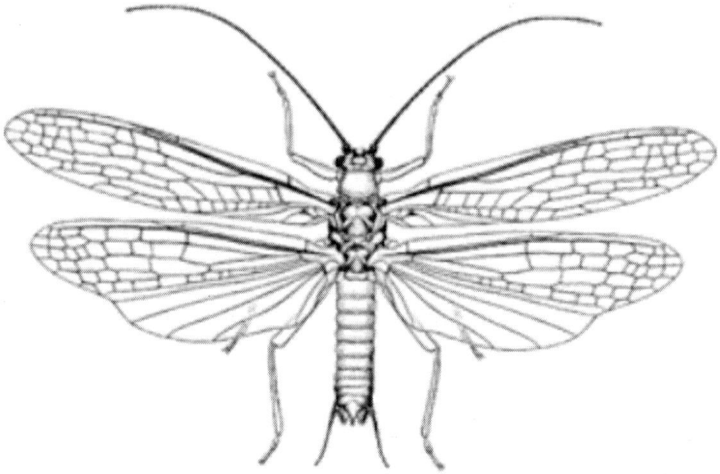

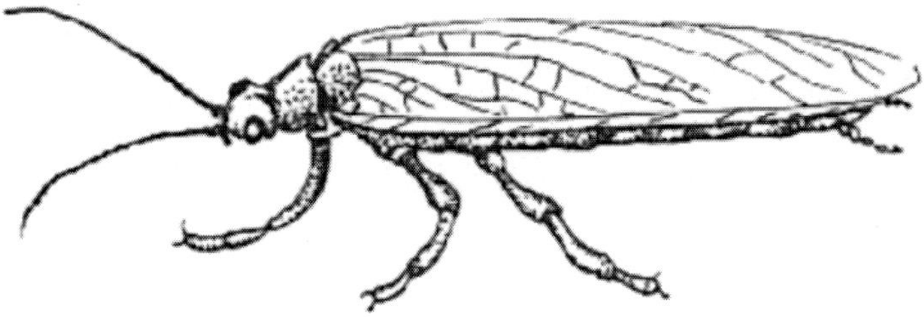

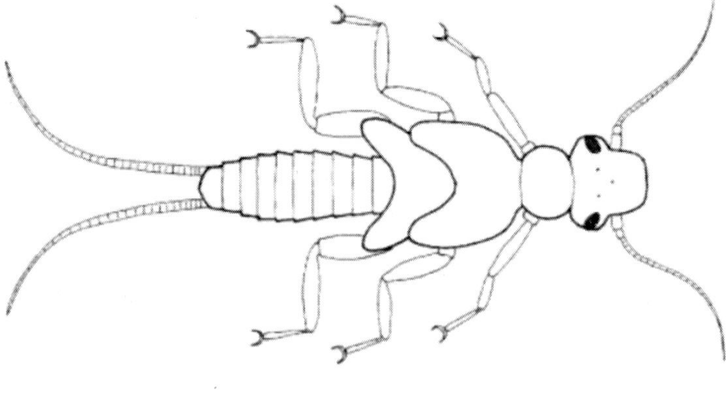

Stonefly

Now we will serve up the Stonefly for all you people who like fishing to help get you excited for the coming season. Stoneflies make up the order of insects called Plecoptera. Worldwide there are 15 families and over 2000 species of stonefly. Nationwide there are 9 families of stonefly, represented by over 600 species. About 120 species live in Montana with about 40 species occurring in the Wolf Creek area. Of these only 2 winter-emerging stoneflies are found in Wolf Creek. These are the Giant Snowfly (Isocapnia *grandis*) and the Slender Snowfly (Capnia *gracilaria*).

These insects emerge in mid to late winter and can often be seen crawling on the snow and ice along the rivers and streams, hence the name "snowfly." They are an important food source to the migrating fish and so an insect of interest to anyone fishing this area from January to April.

Stoneflies, as a group, include snowflies, needleflies, forestflies, willowflies, sallflies, roachflies, springflies and salmonflies. All of these species of flies are very sensitive to water pollution. This means they are a good indicator of environmental health. If stoneflies are present then that stream or river is clean and healthy. Flies from most of these groups are found here in Wolf Creek and all groups are represented in Montana. Talk with our local guides to find out what the fish are biting on at different times of the year.

Stoneflies are seen (and eaten by fish) in two forms; as the **nymph**, and as the adult. The **nymph** varies in size depending on the species, from 1/4 inch to almost an inch in length. It has a head with compound eyes and two well-developed antennae, a 3 segmented ***tarci*** (made up of a **prothorax**, **mesothorax**, and **metathorax**) and an abdomen with two **cerci** that look like tails. It has a pair of legs or claws attached to each of the tarsal segments for a total of 6 legs. The adult resembles the **nymph**, but has a pair of folded wings attached to the **metathorax**.

Stoneflies lay their eggs in masses in streams. The eggs, covered in a sticky substance, sink to the bottom and become attached to rocks or twigs or other debris. The larvae may hatch from the eggs quickly and go through as many as 23 instars as **nymphs**. This stage of the life cycle can last from 1 to 4 years. The mature nymph will climb out of the water for its final molt to an adult. There are a few species, however, that never leave the water; living their entire lives underwater. The adults live from a few days to a few weeks at which time they reproduce and then die.

What am I? Stonefly

What is my Body Type?

Hard ☐ Segmented ☒ How many Segments?

Soft ☒ Worm like ☐ 2 ☐ 3 ☒ More ___

Do I have Legs? Yes ☒ How many? 6 No ☐

Do I have Wings? Yes ☒ How many? 4 No ☐

Are my Wings under ***Elytra***? Yes ☐ No ☒

Do I have Stripes? Yes ☐ No ☒ How Many? ___

Do I have Spots? Yes ☐ No ☒ How Many? ___

What Color am I? green My Spots or Stripes? ______

Kingdom Animalia

Phylum Arthropoda

Class Insecta

Order Plecoptera

Family Capniidae

Genus Isocapnia

Species I. grandis

Draw Me

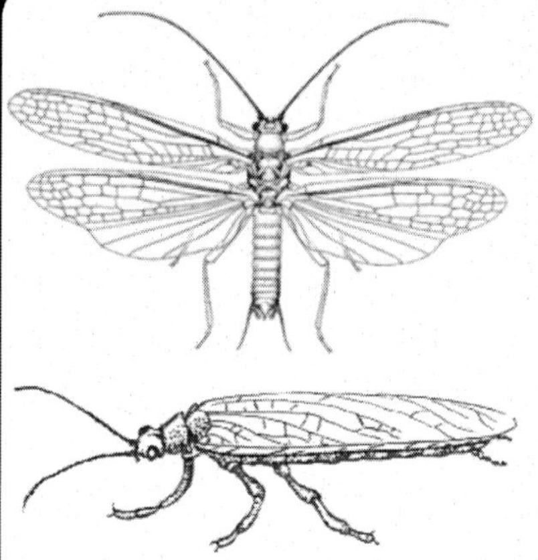

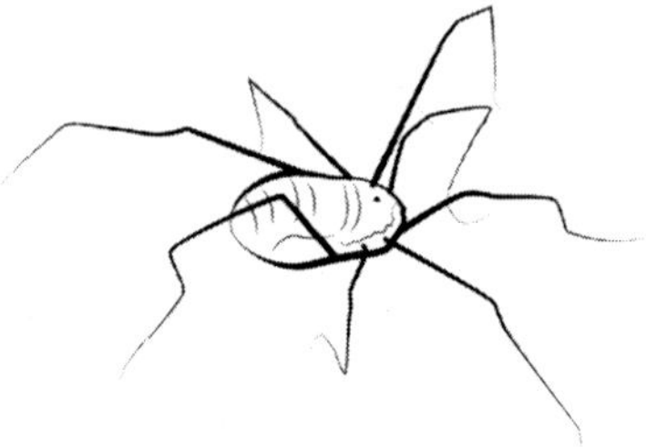

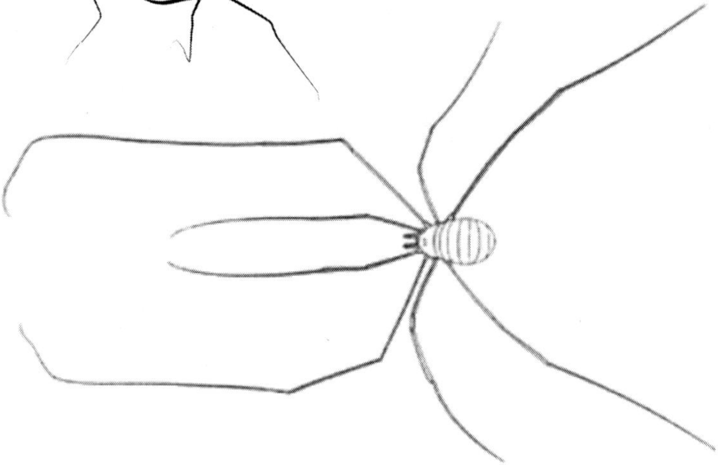

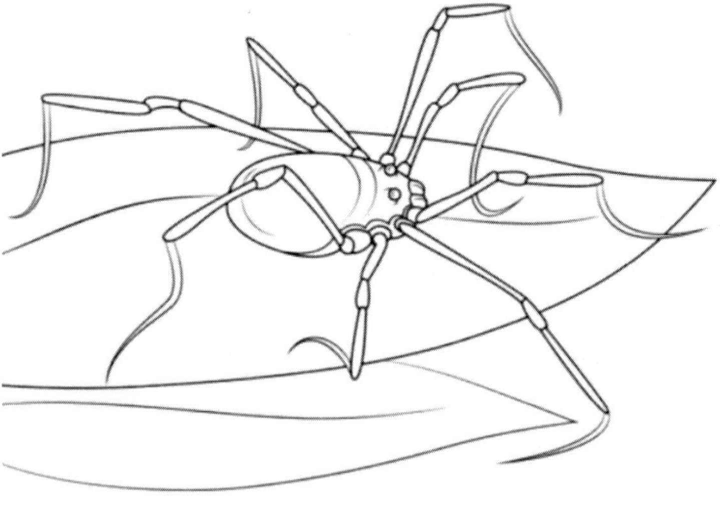

Daddy Long Legs

Harvestman

On the menu here we present an insect that we are all familiar with, but few know very well. It is the Daddy Long Legs, sometimes called the Harvestman. Most people think it is a spider and it even has a myth that follows it around as being the deadliest spider on earth, but has fangs that are too small to bite humans. This tale has been lurking around for years. I heard it a lot when I was growing up and – sad to say – probably spread that rumor around a little myself. It does make a good story to scare your friends with anyway. However, this story is incorrect, but to clarify it, several points need to be explained first.

Most people who retell this tale have no idea that they are referring to 2 completely different groups of bugs: "Daddy Long Legs" and "Daddy Long Legs Spiders." Although the Daddy Long Legs looks like a long-legged spider, it is not a spider. It is one of the Opilones, a group of arachnids closely related to spiders. Unlike the spider, it has no silk glands so is not able to spin a web. It does not have fangs and does not produce venom. It catches its insect prey by using hooks on the ends of its legs. These arachnids defend themselves by secreting a foul-smelling fluid. If caught, they can shed a leg to escape.

The creatures most correctly called Daddy Long Legs are in their own separate Order which is Opiliones and are also called Harvestmen. They are characterized by having one basic body segment which shows segmentation on the posterior portion, at most 2 eyes and all 8 legs attached to the pill-like body segment. The Harvestman is very distinctive with a round, compact body and extremely long legs.

Opilionids most often have long flexible legs in the temperate Northern hemisphere but there are also short-legged Daddy Long Legs. Since they do not produce silk, they are never found in webs unless they are being eaten by spiders. Opilionids are usually found under logs and rocks. They prefer moist habitats although they can be found in the desert. They are often found in gardens, but are sometimes found in the house, especially in basements. They are most often seen in the fall.

These arachnids eat decomposing vegetative and

animal matter. They are opportunistic predators and if they can get away with it, they will eat small insects, snails, and worms. They do not have venom glands, fangs, or any other mechanism for chemically subduing their food. Therefore, they do not have venom. Some have defensive secretions that might be poisonous to small animals if ingested. So, for these daddy-long-legs, the tale of being the "deadliest spider on earth" is clearly false.

What am I? Daddy Long Legs

What is my Body Type?

Hard [X] Segmented [X] How many Segments?

Soft [ ] Worm like [ ] 2 [ ] 3 [ ] ~~More~~ 1

Do I have Legs? Yes [X] How many? 8 No [ ]

Do I have Wings? Yes [ ] How many? ______ No [X]

Are my Wings under ***Elytra***? Yes [ ] No [X]

Do I have Stripes? Yes [ ] No [X] How Many? ____

Do I have Spots? Yes [ ] No [X] How Many? ____

What Color am I? brown My Spots or Stripes? ______

Kingdom Animalia

Phylum Arthropoda

Class Arachnida

Order Opiliones

Family Leiobunidae

Genus Leiobunum

Species many

Draw Me

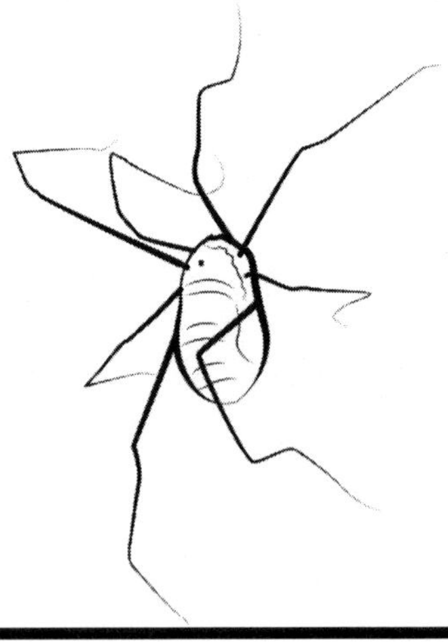

What am I? ______________

What is my Body Type?

Hard ☐ Segmented ☐ How many Segments?

Soft ☐ Worm like ☐ 2 ☐ 3 ☐ More ___

Do I have Legs? Yes ☐ How many? ______ No ☐

Do I have Wings? Yes ☐ How many? ______ No ☐

Are my Wings under ***Elytra***? Yes ☐ No ☐

Do I have Stripes? Yes ☐ No ☐ How Many? ___

Do I have Spots? Yes ☐ No ☐ How Many? ___

What Color am I? ______ My Spots or Stripes? ______

Kingdom __________

Phylum __________

Class __________

Order __________

Family __________

Genus __________

Species __________

Draw Me

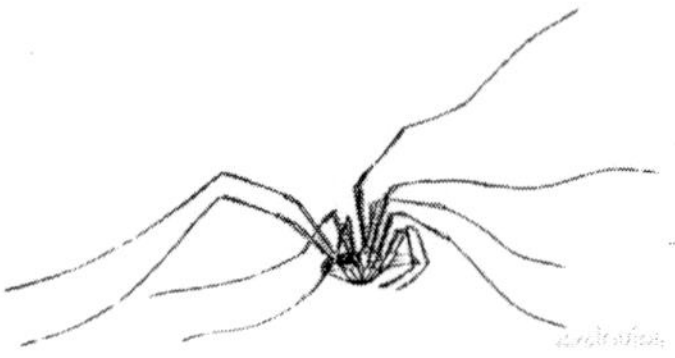

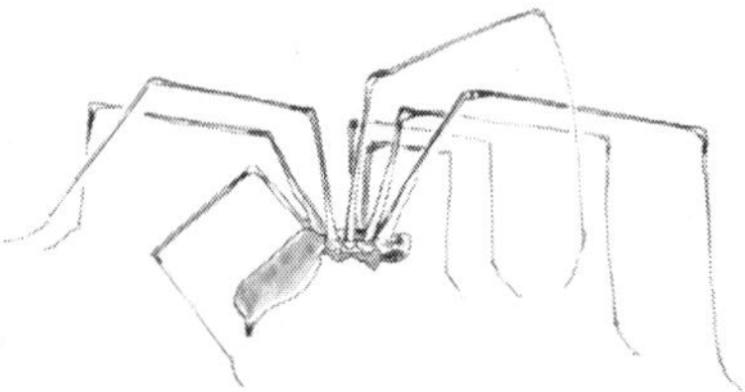

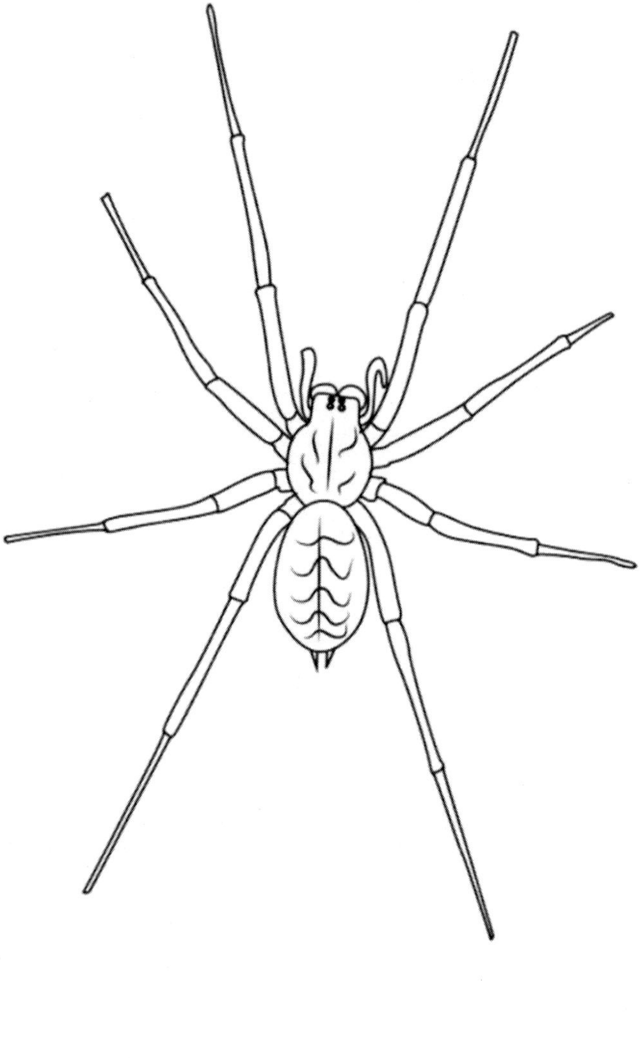

DADDY LONG LEGS

Cellar Spider

Here we serve up another creature often called Daddy Long Legs that is actually a spider. These long-legged spiders are in the family Pholcidae. Previously, the common name of this family was the Cellar Spider but arachnologists – people who study spiders and other arachnids – also call them "Daddy Longleg Spiders" because the general public – people like you and me – call several different bugs “Daddy Longlegs.” Since these arachnids are spiders, they have 2 basic body parts – a **cephalothorax** and an **abdomen**. They have 8 eyes - most often clumped together in the front of the body. Their abdomen shows no evidence of segmentation. They have 8 legs all attached to the front most body part - the **cephalothorax** - and make webs out of silk. This is probably the bug most people refer to when they tell the tale of Daddy Longlegs being the deadliest spider on earth – I don’t know though, when I was a kid in the Pacific Northwest, what we called a Daddy Long Leg was really a Harvestman.

Pholcid spiders are plentiful especially in cellars - hence their other common name - and are commonly seen by the general public more often than the Harvestman. The two most common pholcid spiders found in U.S. homes are both European immigrants. Pholcus *phalangioides* is a uniformly grey spider with a rectangular, elongate abdomen and is found throughout all North America. Holocnemus *pluchei* also has a rectangular, elongate **abdomen**. It has a brown stripe on the ventral side - the belly side – of its **abdomen**. The stripe is typically directed upwards since the spider hangs upside down in its web. These spiders are very common along the Pacific Coast. and into the southwest deserts.

Cellar spiders spin loose, irregular webs in homes, sheds, barns and warehouses. Their webs are usually found in dark and damp places like the corners of eaves, windows, and ceilings in cellars, basements, crawlspaces, and garages.

Cellar spiders prefer to hang upside down in their webs as they wait for prey. They eat other spiders and insects. When bothered, a cellar spider will repeatedly shake its body to make its entire web pulse. These pulsations help to entrap any insects that have approached the web and will soon become the spider's next meal.

Here, the myth about this bug being the "deadliest spider on earth" is incorrect at least in making claims that have no basis in known facts. There is no reference to any pholcid spider biting a human and causing any bad reaction. Furthermore, there are no toxicological studies testing the lethality of pholcid venom on any mammalian system - this is usually done with mice. Therefore, no information is available on the toxic effects of their venom in humans, so the part of the myth about their being especially poisonous is just that: a myth. There is no scientific basis for the idea that they are deadly poisonous and there is no reason to assume that it is true.

What about their fangs being too short to penetrate human skin? Pholcids do indeed have short fangs. In arachnological terms this is called "*uncate*." They have secondary teeth which meet the fangs just like a pair of tongs. Brown recluse spiders similarly have an uncate fang structure and they can bite humans. There may be differences in the muscle structure that works the fangs giving brown recluses stronger muscles for biting prey. Recluses are hunting spiders that need to subdue their prey, whereas pholcid spiders are able to wrap their prey in spider silk and don't need as strong a bite to catch their prey. There is no scientific basis for the myth that either Harvestmen or Cellar Spiders are the "deadliest spiders on earth" so, please, stop spreading that story.

What am I? Daddy Long Legs

What is my Body Type?

Hard [X] Segmented [X] How many Segments?

Soft [ ] Worm like [ ] 2 [X] 3 [ ] More ___

Do I have Legs? Yes [X] How many? 8 No [ ]

Do I have Wings? Yes [ ] How many? ___ No [X]

Are my Wings under ***Elytra***? Yes [ ] No [X]

Do I have Stripes? Yes [ ] No [X] How Many? ___

Do I have Spots? Yes [X] No [ ] How Many? 5

What Color am I? tan My Spots or Stripes? brown

Kingdom Animalia

Phylum Arthropoda

Class Arachnida

Order Araneae

Family Pholcidae

Genus Pholcus

Species phalangioides

Draw Me

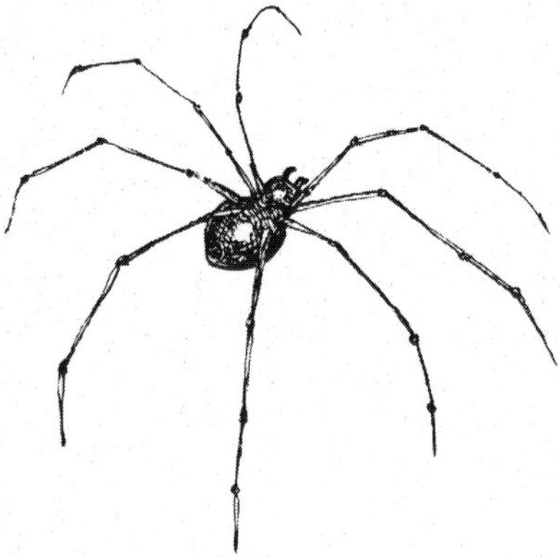

What am I? ____________

What is my Body Type?

Hard ☐ Segmented ☐ How many Segments?

Soft ☐ Worm like ☐ 2 ☐ 3 ☐ More ____

Do I have Legs? Yes ☐ How many? ______ No ☐

Do I have Wings? Yes ☐ How many? ______ No ☐

Are my Wings under ***Elytra***? Yes ☐ No ☐

Do I have Stripes? Yes ☐ No ☐ How Many? ____

Do I have Spots? Yes ☐ No ☐ How Many? ____

What Color am I? ______ My Spots or Stripes? ______

Kingdom ________

Phylum ________

Class ________

Order ________

Family ________

Genus ________

Species ________

Draw Me

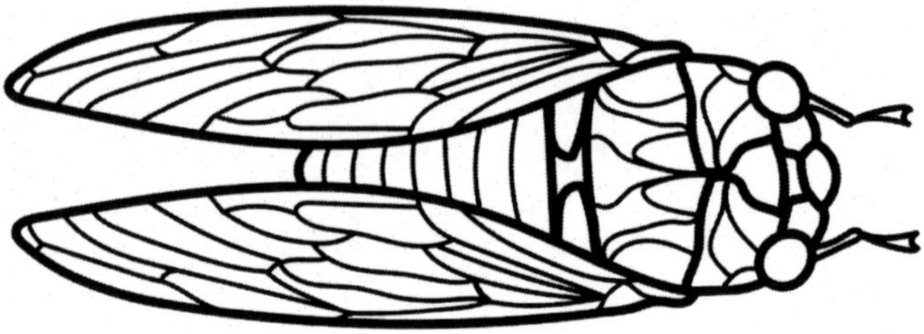

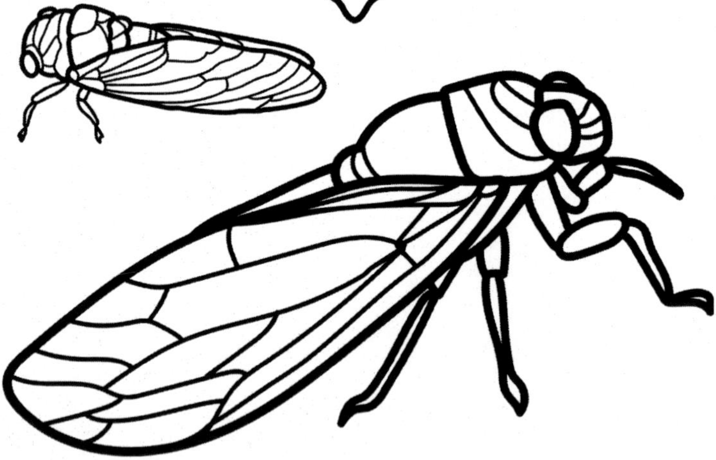

Cicada

On the platter today is a noisy dish known as the Cicada, brought to our attention by Rocky Infanger and David Keitel who found them at the Log Gulch campground on Holter lake. There are several different species of cicada belonging to at least three different genera.

First there is the Periodical Cicada, sometimes called the 13-year or 17-year Cicada and scientifically known as Magicicada *tredecim* or Magicicada *septendecim*. These are what some people call "17-year Locusts" but they are not a locust at all. Locusts are grasshopper-type insects. There are also Annual Cicadas that emerge every spring and are commonly known as Dog Day Cicadas, Harvest Flies, or by their color such as Green Annual, Brown Annual or Golden Annual Cicadas, belonging to the genus Tibicen, and there is the Pygmy Cicada (Neocicada *heiroglyphica*).

The most noticeable thing about cicadas is their song, or noise, that loud incessant buzzing - that gets louder (and louder and LOUDER - Stop! please!

No more! I can't take it anymore!) the warmer it gets. Why do they make that noise anyway? It is the males (wouldn't you know it?) trying to attract females, but it attracts other males too because they like to make noise together (imagine that!). Their "song" varies by species and has been described as sounding like anything from a "small car engine trying to start" to "much like a circular table saw cutting through a wooden plank." Each insect's individual song lasts 10 to 20 seconds and is repeated every few seconds. When you put several hundreds, or millions even, together imagine the noise they make.

Periodical Cicadas are the longest living insects in North America and are native to this continent. There are 3 known species of 17-year and 3 known species of 13-year cicada. This insect naturally occurs nowhere else in the world. They emerge from the ground in late spring as the weather gets drier. They mate over the next ten days. After mating, females lay groups of 24 to 28 eggs in cuts they have made on branches. They may lay as many as 600 eggs before they die. The adults live 3 to 4 weeks above ground. The **nymphs** hatch out of the eggs after 6 to 7 weeks. The hatchlings drop to the ground and burrow down to 2 feet and feed off the plant juices in roots. After 13 or 17 years the **instars** leave the ground, climb up trees, bushes or weeds then shed their nymphal skin emerging as adults, ready to start the next cycle. The adults are about 1 and a 1/2 inches long and reddish brown with red to orange eyes. Their wings are larger than their bodies and transparent with reddish veins. Annual cicadas are larger, up to 2 inches, and pygmy cicadas only grow to 3/4 of an inch.

What am I? Cicada

What is my Body Type?

Hard [X] Segmented [X] How many Segments?

Soft [ ] Worm like [ ] 2 [ ] 3 [X] More ____

Do I have Legs? Yes [X] How many? 6 No [ ]

Do I have Wings? Yes [X] How many? 4 No [ ]

Are my Wings under ***Elytra***? Yes [ ] No [X]

Do I have Stripes? Yes [ ] No [X] How Many? ____

Do I have Spots? Yes [ ] No [X] How Many? ____

What Color am I? black My Spots or Stripes? ______

Kingdom Animalia

Phylum Arthropoda

Class Insecta

Order Hemiptera

Family Cicadidae

Genus Magicicada

Species tredecim

Draw Me

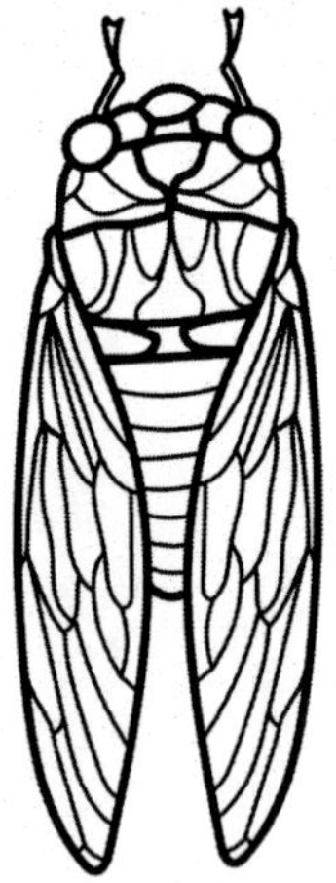

What am I? ____________

What is my Body Type?

Hard ☐ Segmented ☐ How many Segments?

Soft ☐ Worm like ☐ 2 ☐ 3 ☐ More ___

Do I have Legs? Yes ☐ How many? ______ No ☐

Do I have Wings? Yes ☐ How many? ______ No ☐

Are my Wings under ***Elytra***? Yes ☐ No ☐

Do I have Stripes? Yes ☐ No ☐ How Many? ____

Do I have Spots? Yes ☐ No ☐ How Many? ____

What Color am I? ______ My Spots or Stripes? ______

Kingdom __________

Phylum __________

Class __________

Order __________

Family __________

Genus __________

Species __________

Draw Me

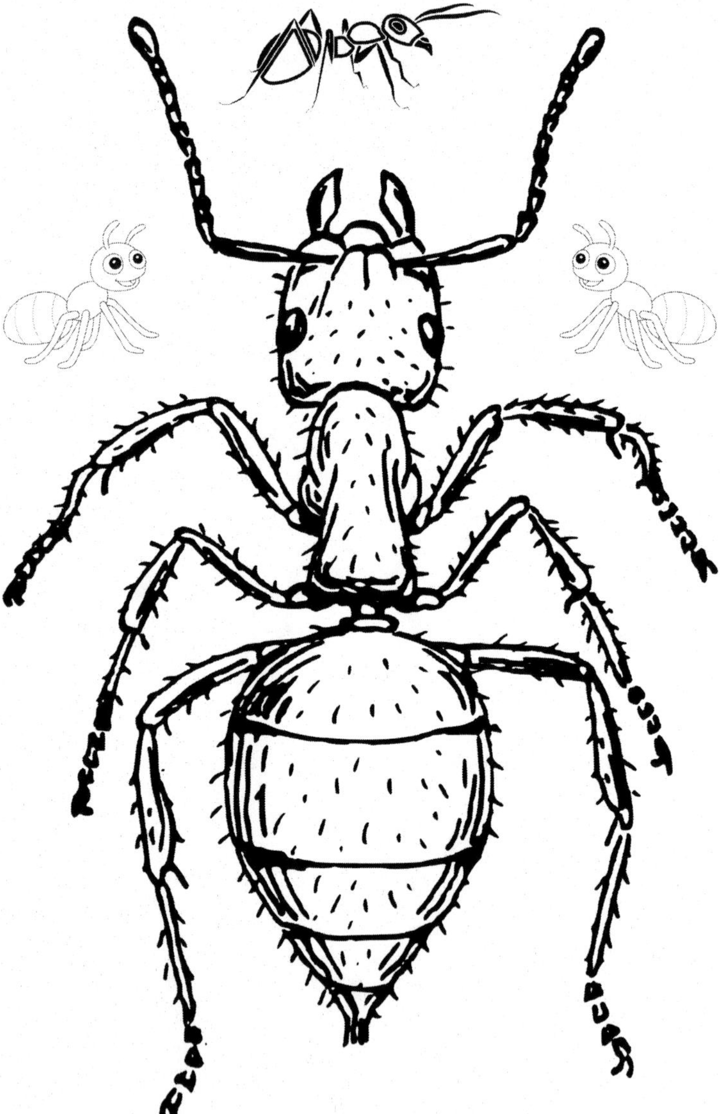

Harvester Ant

Now we offer up the Western Harvester Ant, entomologically known as Pogonomyrmex *occidentalis* and sometimes erroneously referred to as the Fire Ant because of its reddish color and painful bite.

Harvester ants live in colonies. Colonies consist of a queen, workers, and **larva**. A colony is started by an impregnated female ant coming from a pre-existing colony. The young queen finds a suitable spot to start her new colony. She tunnels into the ground and starts a brood chamber where she lays her eggs. The young queen cares for these first eggs herself. They will develop into sterile females called workers. The mature workers will take care of the queen and all the newly laid eggs as well as tend to the nest and forage for food.

Periodically, usually after a summer rain, some of the eggs will develop into fertile females (young queens) and fertile males (drones), both of which have wings. These flying ants leave the colony and "swarm" to reproduce. The female can be impregnated by several males and will store the sperm for the rest of her life to fertilize her eggs as she lays them (she is fertilized only once in her lifetime). The newly impregnated female will start a new colony of her own. The new queen may live for ten to twenty years (that is not a misprint, I did say 20 years!). The drones live only a few days whether or not they successfully reproduce.

Workers are the most visible members of the Harvester Ant colony. They are three-eighths to a half inch in length. they are reddish-brown to black in color. They have squarish heads with large mandibles. Their bodies are covered with short hairs, but you have to get really close to see that.

Harvester ants build their nests in the ground and are usually easy to spot because there is a large mound at the entrance. There are tunnels and chambers in the mound as well as underground. Nests can be quite extensive, that is they can decend into the ground a foot or more and spread out several feet from the mound.

These ants got their name because they harvest

the seeds of plants for food. They usually go after one type of seed until that is no longer available then they collect another variety. Workers may forage as far as a hundred and fifty feet away from their mound. Because of their foraging habits, these ants do play a decisive role in what type of vegetation is growing in the areas they inhabit.

Harvester ants have a strong venom and do bite, although they are not normally aggressive unless their nest is being disturbed. The bite is painful and can cause an allergic reaction, so it should be tended to as soon as possible.

What am I? Harvester Ant

What is my Body Type?

Hard [X] Segmented [X] How many Segments?

Soft [ ] Worm like [ ] 2 [ ] 3 [X] More ____

Do I have Legs? Yes [X] How many? 6 No [ ]

Do I have Wings? Yes [ ] How many? ______ No [X]

Are my Wings under ***Elytra***? Yes [ ] No [X]

Do I have Stripes? Yes [ ] No [X] How Many? ____

Do I have Spots? Yes [ ] No [X] How Many? ____

What Color am I? red My Spots or Stripes? ______

Kingdom Animalia

Phylum Arthropoda

Class Insecta

Order Hymenoptera

Family Formicidae

Genus Pogonomyrmex

Species occidentalis

Draw Me

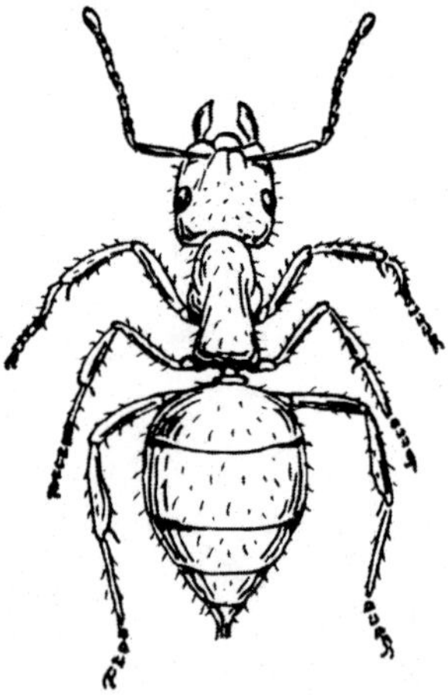

What am I? ____________

What is my Body Type?

Hard ☐ Segmented ☐ How many Segments?

Soft ☐ Worm like ☐ 2 ☐ 3 ☐ More ___

Do I have Legs? Yes ☐ How many? ______ No ☐

Do I have Wings? Yes ☐ How many? ______ No ☐

Are my Wings under ***Elytra***? Yes ☐ No ☐

Do I have Stripes? Yes ☐ No ☐ How Many? ___

Do I have Spots? Yes ☐ No ☐ How Many? ___

What Color am I? ______ My Spots or Stripes? ______

Kingdom ________

Phylum ________

Class ________

Order ________

Family ________

Genus ________

Species ________

Draw Me

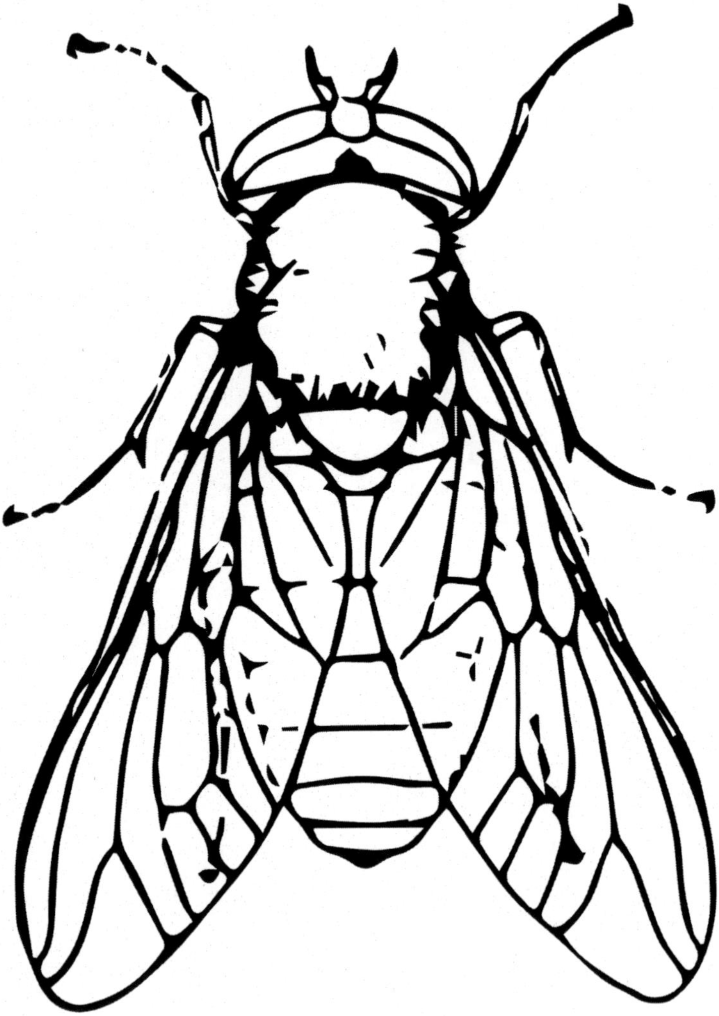

Fruit Fly

This time we are serving an appetizer, or in the very least, a very small bug, the Fruit Fly. In fact, it is one of the smallest flies found in homes and around humans. There are many species of fruit fly, but the most commonly encountered is Drosophila *melanogaster*. Fruit flies are also known as Pomace Flies or Vinegar Flies. They are those little insects that hover around fruit and vegetables that have been out for several days. They can be found hanging around those empty beer bottles and pop cans by the garbage or the recycle bin too.

"Where do they come from?" you may ask yourself. Well, in the 19th century it was generally believed that they "spontaneously generated" out of fruit matter or left-over beer and wine, but we now know that is not so. Fruit flies live everywhere in the world (except in the extremely cold parts). They live off decaying plant matter and are attracted to anything that is fermenting, like overripe fruit, old vegetables, left over beer, wine, or pop, and even damp cotton (things like old rags, t-shirts, mops, etc.).

The females lay their eggs on any damp,

fermenting surface (fruit, drain slime, garbage, etc.) and the eggs hatch in a day or two. The **larva** (maggot) feeds on the fermenting substance for four or five days then crawls away to drier areas to pupate. The complete life cycle averages ten days, but can be as little as 8 or as many as fourteen. A single female can lay up to five hundred eggs.

Because these insects are surface feeders, the bad portions of the infested fruit can be cut away and the good parts still eaten, but ingesting the little maggots can cause intestinal pain and cramping along with diarrhea. Also, because of the mobility of the adults and their choice of feeding and breeding areas, they have the potential to carry and spread disease-causing bacteria and other germs.

Adult fruit flies are less than eighth of an inch long. They appear black, but are actually tan to golden brown with black stripes around the back part of their **abdomen**. They usually have red eyes. They have one pair of wings and six legs.

Their short life cycle and abundance of egg-laying makes them perfect specimens for scientific study - especially in the field of genetics. They are relatively easy and inexpensive to keep and culture so they are very popular for biology classes.

They are also fairly easy to get rid of and keep rid of, generally by not letting fruit or vegetables sit around uncovered for long periods of time, keeping the garbage covered, keeping drains cleaned and drying out your mops and cleaning rags.

There is all kinds of information on these things, they are probably the most studied insect in the world, so have a little fun and see what you can learn.

What am I? Fruit Fly

What is my Body Type?

Hard [X] Segmented [X] How many Segments?

Soft [ ] Worm like [ ] 2 [ ] 3 [X] More ____

Do I have Legs? Yes [X] How many? 6 No [ ]

Do I have Wings? Yes [X] How many? 2 No [ ]

Are my Wings under ***Elytra***? Yes [ ] No [X]

Do I have Stripes? Yes [X] No [ ] How Many? 5

Do I have Spots? Yes [ ] No [X] How Many? ____

What Color am I? brown My Spots or Stripes? black

Kingdom Animalia

Phylum Arthropoda

Class Insecta

Order Diptera

Family Drosophilidae

Genus Drosophila

Species melanogaster

Draw Me

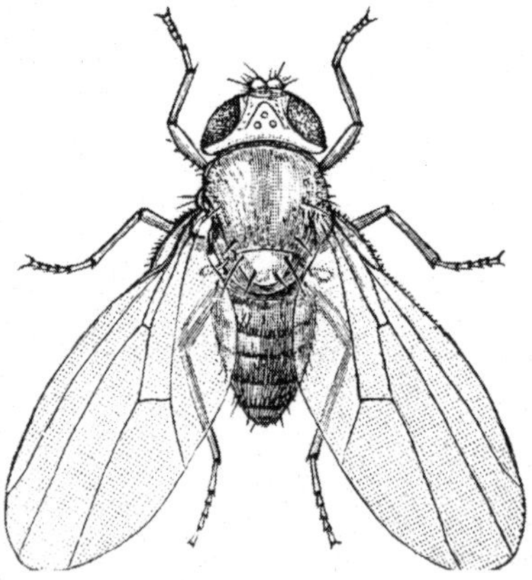

What am I? ____________

What is my Body Type?

Hard ☐ Segmented ☐ How many Segments?

Soft ☐ Worm like ☐ 2 ☐ 3 ☐ More ___

Do I have Legs? Yes ☐ How many? ______ No ☐

Do I have Wings? Yes ☐ How many? ______ No ☐

Are my Wings under ***Elytra***? Yes ☐ No ☐

Do I have Stripes? Yes ☐ No ☐ How Many? ___

Do I have Spots? Yes ☐ No ☐ How Many? ___

What Color am I? ______ My Spots or Stripes? ______

Kingdom ________

Phylum ________

Class ________

Order ________

Family ________

Genus ________

Species ________

Draw Me

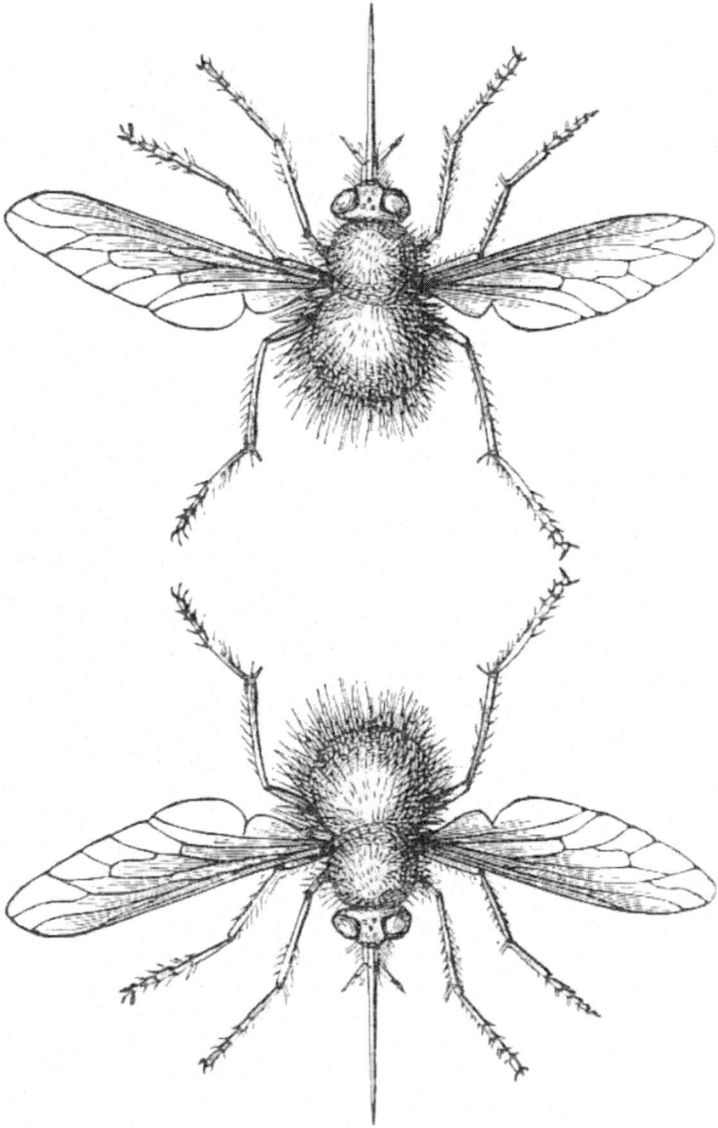

BEE FLY

Here we treat you to an insect brought to our attention by our local Deputy Sheriff, Allen Hughes, and our local Fire Chief, Rocky Infanger. It is the Bee Fly. The specimen I was brought is Systoechus *oreas*, one of five species in the genus Systoechus of the family Bombyliidae.

The Bee Fly mimics bees in behaviour and appearance, thereby benefiting from the characteristics of bees, hence its name. Potential predators associate bee stings with insects that look and act like bees so the predators tend to leave those insects alone.

The bee fly, however, only looks and sounds like a bee. It cannot sting or bite. Adult bee flies are one quarter-inch to one half-inch long and have stout bodies covered in hair. They have a long **proboscis**, like butterflies, that they use to get nectar from flowers. Many species' bodies are vividly colored brown to black and yellow, resembling bee coloration. The wings are usually transparent, but may be dark colored on the front half with only the rear half being transparent. They buzz loudly, like bees, especially when caught in nets or trapped.

Adult bee flies feed on nectar and pollen and hang

around flowers just like real bees. The adults can be found from early May to late August. The females lay their eggs in the nests of ground-nesting insects such as bees, wasps, moths, grasshoppers and beetles. The maggots (**larvae**) of bee flies feed on the larvae and pupae of those insects, making them beneficial in the control of harmful insect species, although they prey on beneficial insect **larvae** too.

There are some seven hundred and fifty species of Bee Fly in forty-three Genera making up the family Bombyliidae on the American continent north of Mexico. Bee Flies seem to be more populous in the drier regions of the western United States than in the more humid regions of the south east. There are also several more families of fly closely related to, and sometimes confused with, bee flies. These include hover flies and flower flies (family Syrphidae) and robber flies (family Asilidae). Together they represent close to three thousand species of bee-like and wasp-like flies. All of these are "true flies" meaning they are in the Order Diptera, in which there are over 120,000 known species or roughly one-third of all known animal species (including spiders, worms, birds, fish and all) on the planet. Oh, my!

What am I? Common Bee Fly

What is my Body Type?

Hard [ ] Segmented [X] How many Segments?

Soft [X] Worm like [ ] 2 [ ] 3 [X] More ___

Do I have Legs? Yes [X] How many? 6 No [ ]

Do I have Wings? Yes [X] How many? 2 No [ ]

Are my Wings under ***Elytra***? Yes [ ] No [X]

Do I have Stripes? Yes [ ] No [X] How Many? ___

Do I have Spots? Yes [ ] No [X] How Many? ___

What Color am I? yellow My Spots or Stripes? ___

Kingdom Animalia

Phylum Arthropoda

Class Insecta

Order Diptera

Family Bombyliidae

Genus Systoechus

Species S. vulgaris

Draw Me

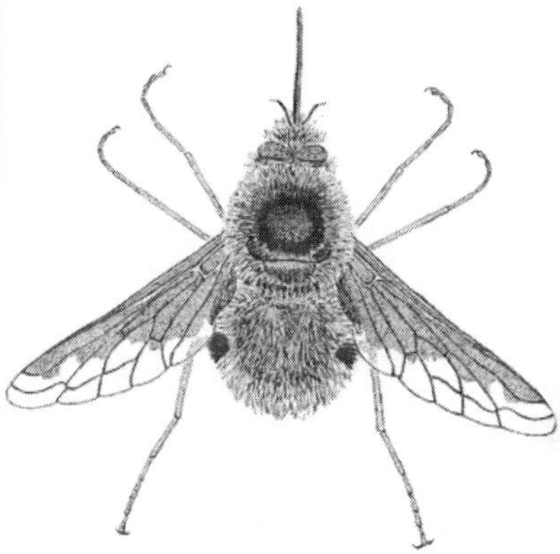

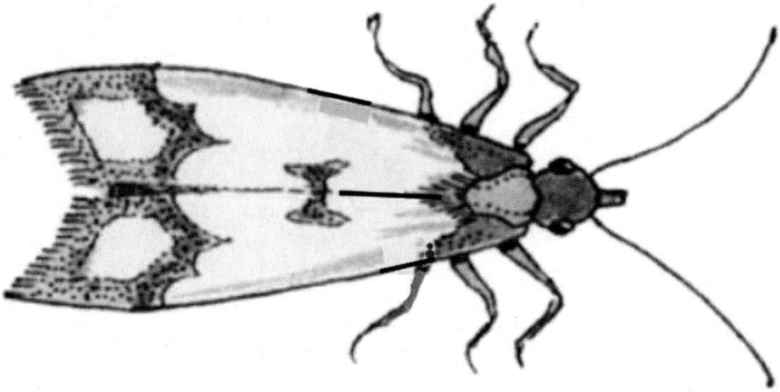

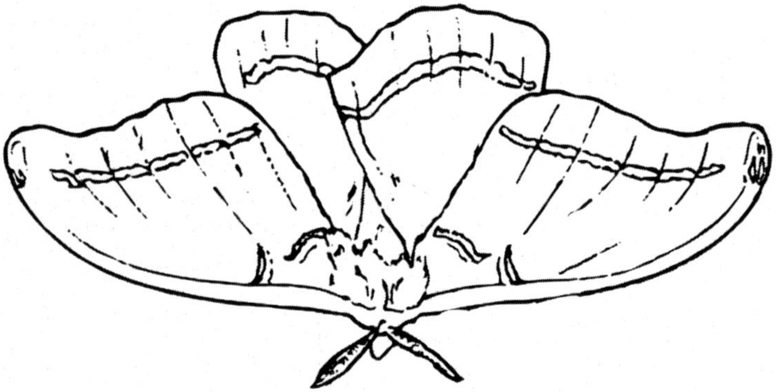

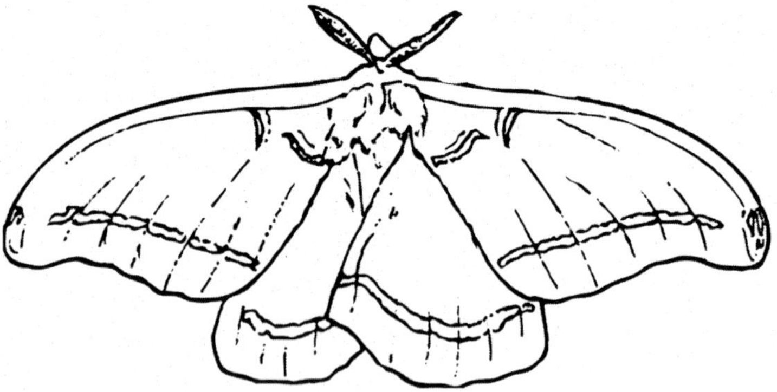

Sulfur Knapweed Root Moth

The special of the day here is the Sulfur Knapweed Root Moth, also known as the Yellow-winged Knapweed Root Moth. For the entomologists out there, it is the Agapeta *zoegana* of the order Lepidoptera, family Torticidae.

This moth has been imported to the United States by the government in an effort to control the infestation of Spotted Knapweed. Spotted Knapweed is not native to the U.S. and as such has no natural enemies so the environmentalists have decided that it would be prudent to establish populations of natural enemies to help the ranchers, farmers and land owners control the weed. Hmmmm. This moth was introduced to Montana in 1984 (!?!) and now has established populations in Montana, Wyoming, Idaho, Washington, Oregon, California, Colorado, Arizona, Utah, Nevada, Nebraska, Minnesota, Wisconsin and both Dakotas.

We are told, by government-paid biologists and entomologists, that this moth only attacks the roots of spotted knapweed and diffuse knapweed. They say it has been tested on 51 different plant species, including safflower and other economically important plants, but only attacks the 2 species of knapweed previously mentioned. Its method of attack is as follows: eggs are laid on the stems and leaves of the host plant. When the larvae hatch, they move to the crown area and eat the roots. This damage will kill the young plants and retard the growth and development of older plants, preventing them from flowering or producing seeds.

The adult moth has bright yellow wings with brown bands towards the bottom of the wing. The adult moth is small - about three-eighths of an inch long. They look like dying knapweed leaves and can be found resting on knapweed stems.

The adults emerge from June to early September. The female mates within the first 24 hours and lays eggs the following day. The female can lay up to 78 eggs in her lifetime. Eggs are laid white then turn yellowish-red after 3 days. The eggs hatch within 7 to 10 days and the **larvae** move directly to the root crown and start devouring the roots. **Larvae** go through 6 **instars**, each stage slightly larger than the previous one. **Larvae** can move from plant to plant if neighboring plants are within 4 inches of each other. The **larvae pupate** in the root. The Adults are strong fliers and can move from one plant population to another. Adults only live 11 to 14 days. The insect over winters in the larval stage.

What am I? Sulfur Knapweed Root Moth

What is my Body Type?

Hard ☐ Segmented ☒ How many Segments?

Soft ☒ Worm like ☐ 2 ☐ 3 ☒ More ___

Do I have Legs? Yes ☒ How many? 6 No ☐

Do I have Wings? Yes ☒ How many? 4 No ☐

Are my Wings under ***Elytra***? Yes ☐ No ☒

Do I have Stripes? Yes ☒ No ☐ How Many? 1

Do I have Spots? Yes ☒ No ☐ How Many? 2

What Color am I? yellow My Spots or Stripes? brown

Kingdom Animalia

Phylum Arthropoda

Class Insecta

Order Lepidoptera

Family Tortricidae

Genus Agapeta

Species A. zoegana

Draw Me

Hummingbird Moth

At the request of Cindy Humphrey, we serve up the hummingbird moth. Several species of the genus *Hemaris* deserve this name because they fly and move just like hummingbirds. They can remain suspended in the air in front of a flower while they unfurl their long tongues and insert them in flowers to sip their nectar. They even sound like hummingbirds. They beat their wings about 70 times per second and can fly up to 12 miles per hour. If you are not looking closely, you may mistake this moth for a hummingbird.

Hummingbird moths have plump bodies and the tips of their tails open into a fan. They can be reddish-brown in color with a yellow and black abdomen, or they can be all black and white striped. Their bodies are 2 to 2 and a half inches long and their wingspan is 6 to 7 inches. Like all Lepidoptera their wings are covered by scales; some species lose many of the scales from patches on their wings, so they are called clearwing hummingbird moths. They have a very long **proboscis** which they use to reach the nectar of long-necked flowers. Their tongue-like **proboscis** can be twice the length of their body! When they are not feeding on nectar, they keep their tongues rolled up in a coil at the front of their heads. Since they feed on flowers that open up during the day, these moths fly and feed when the sun is out.

There are four species of hummingbird moths in North America. The most familiar ones are the Snowberry Clearwing (Hemaris *diffinis*) and the Hummingbird Clearwing (Hemaris *thysbe*). They are both widespread throughout North America. The Snowberry Clearwing is more abundant in the west and the Hummingbird Clearwing is more common in the east. The other two species are the Slender Clearwing Sphinx moth (Hemaris *gracilis*) and the White-lined Sphinx moth (Hyles *lineata*). The Sphinx moth is active both day and night. The flowers that bloom after dark; like the moonflower, attract them. The sphinx moths are sometimes called Bee Hawk Moths – especially in Great Britain. Like most moths and butterflies, the adult hummingbird moths feed on nectar from a variety of flowers. Their **larvae** need more specific food plants; such as several species of honeysuckle, dogbane, hawthorn, cherries, and plums.

The adults usually start flying in early spring, but you will have a better chance of seeing them in the summer when they are most active. The female Hummingbird Moth attracts the males with a **pheromone** – smells - which they produce from glands at the tip of their abdomen. After mating, the females lay their eggs on the underside of the leaves of plants that their **larvae** like to eat.

The caterpillars have a horn at the rear end and are commonly green to blend in with the leaves. When the **larvae** are fully-grown they drop to the ground, crawl under the dead leaves and spin a loose cocoon to **pupate**. In the north, where the season is short there is only one generation per year. The **pupa** remains hidden under the leaf litter all winter and the adult moth does not emerge until the following spring. In the south, where it is warmer longer, there is usually more than one generation each summer.

What am I? hummingbird moth

What is my Body Type?

Hard ☐ Segmented ☒ How many Segments?

Soft ☒ Worm like ☐ 2 ☐ 3 ☒ More ____

Do I have Legs? Yes ☒ How many? 6 No ☐

Do I have Wings? Yes ☒ How many? 4 No ☐

Are my Wings under ***Elytra***? Yes ☐ No ☒

Do I have Stripes? Yes ☒ No ☐ How Many? 4

Do I have Spots? Yes ☐ No ☒ How Many? ____

What Color am I? yellow My Spots or Stripes? black

Kingdom Animalia

Phylum Arthropoda

Class Insecta

Order Lepidoptera

Family Sphingidae

Genus Hemaris

Species H. diffinis

Draw Me

What am I? ____________

What is my Body Type?

Hard ☐ Segmented ☐ How many Segments?

Soft ☐ Worm like ☐ 2 ☐ 3 ☐ More ___

Do I have Legs? Yes ☐ How many? ______ No ☐

Do I have Wings? Yes ☐ How many? ______ No ☐

Are my Wings under ***Elytra***? Yes ☐ No ☐

Do I have Stripes? Yes ☐ No ☐ How Many? ____

Do I have Spots? Yes ☐ No ☐ How Many? ____

What Color am I? ______ My Spots or Stripes? ______

Kingdom ________

Phylum ________

Class ________

Order ________

Family ________

Genus ________

Species ________

Draw Me

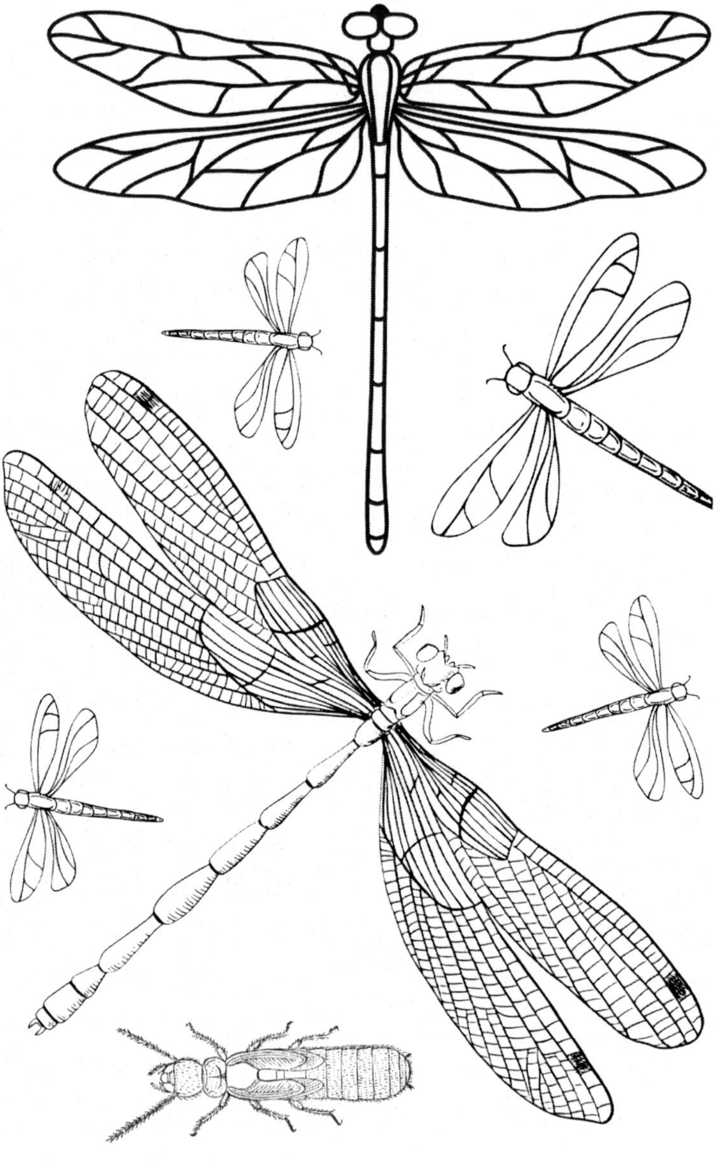

Dragonfly

Here I'd like to offer up the Dragonfly. It is, perhaps, my own personal favorite insect. Dragonflies are found on every continent except Antarctica – like many other insects. There are 11 families divided into 348 genera with over 3,000 species of dragonfly in the world. Most of them are found in tropical areas, but there are about 450 species in North America. The Common Green Darner (Anax *junius*) is the most widespread and most frequently seen large dragonfly in the United States. It is about 3 inches long with a wingspan of just over 3 inches. It has a green thorax and a turquoise-blue **abdomen**. This is the official state insect of Washington.

Dragonflies have large bodies – divided into three segments like all insects. They have 2 pair of wings that extend horizontally out from their bodies while in flight and at rest. They have a large head with short antennae and large **compound** eyes. Their **thorax** is round and bulbous and is where all 4 wings are attached. Their **abdomen** is long and thin. All these attributes together are what give the dragonfly its distinctive look.

Adult dragonflies hunt while they are flying. They eat a wide variety of small insects like midges, mosquitoes, damselflies, and even smaller dragonflies. They eat their prey headfirst, but don't eat the wings. A dragonfly may consume as much as a fifth of its body weight in prey per day. Dragonflies are some of the insect world's most efficient hunters, catching up to 95% of the prey they go after. They can fly at speeds of almost 10 mph, some of the bigger species can fly as fast as 34 mph. The **nymphs** are voracious predators, eating most living things that are smaller than they are. Their diet mostly consists of insect **larvae,** but they also feed on tadpoles and small fish.

Dragonflies have a complex mating ritual and the eggs are usually laid while the male and female are still holding on to each other. Eggs are laid in different ways, depending on the species, but always on vegetation in or near water. They may lay as many as 1500 eggs at a time. The aquatic **nymphs** – **naiads** – hatch after about a week. They **molt** 6 to 15 times as they grow towards adulthood, depending on the species. The **nymph** stage lasts up to 5 years in larger species, but only 2 months to 3 years in smaller species.

When a **nymph** is ready to **molt** into an adult it

crawls out of the water on a plant stem then clings onto it with its claws. The **exoskeleton** splits at a weak spot just behind its head then the adult emerges from the old shell. Just before the end of its **abdomen** exits the **exoskeleton**, the dragonfly stops while it gulps air to pump up its wings after which it can fly away, leaving the **exoskeleton** still clasping the plant stem. Adult dragonflies live just a few weeks. They can fly over 900 miles in their short life span!

What am I? darner dragonfly

What is my Body Type?

Hard [X] Segmented [X] How many Segments?

Soft [ ] Worm like [ ] 2 [ ] 3 [X] More ____

Do I have Legs? Yes [X] How many? 6 No [ ]

Do I have Wings? Yes [X] How many? 4 No [ ]

Are my Wings under ***Elytra***? Yes [ ] No [X]

Do I have Stripes? Yes [ ] No [X] How Many? ____

Do I have Spots? Yes [X] No [ ] How Many? 4

What Color am I? green n blue My Spots or Stripes? black

Kingdom Animalia

Phylum Arthropoda

Class Insecta

Order Odonata

Family Aeshnidae

Genus Anax

Species A. junius

Draw Me

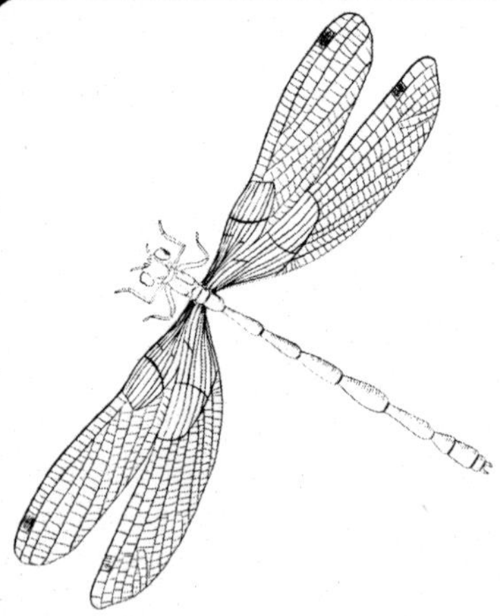

What am I? ________________

What is my Body Type?

Hard ☐ Segmented ☐ How many Segments?

Soft ☐ Worm like ☐ 2 ☐ 3 ☐ More ___

Do I have Legs? Yes ☐ How many? ______ No ☐

Do I have Wings? Yes ☐ How many? ______ No ☐

Are my Wings under ***Elytra***? Yes ☐ No ☐

Do I have Stripes? Yes ☐ No ☐ How Many? ____

Do I have Spots? Yes ☐ No ☐ How Many? ____

What Color am I? ______ My Spots or Stripes? ______

Kingdom __________

Phylum __________

Class __________

Order __________

Family __________

Genus __________

Species __________

Draw Me

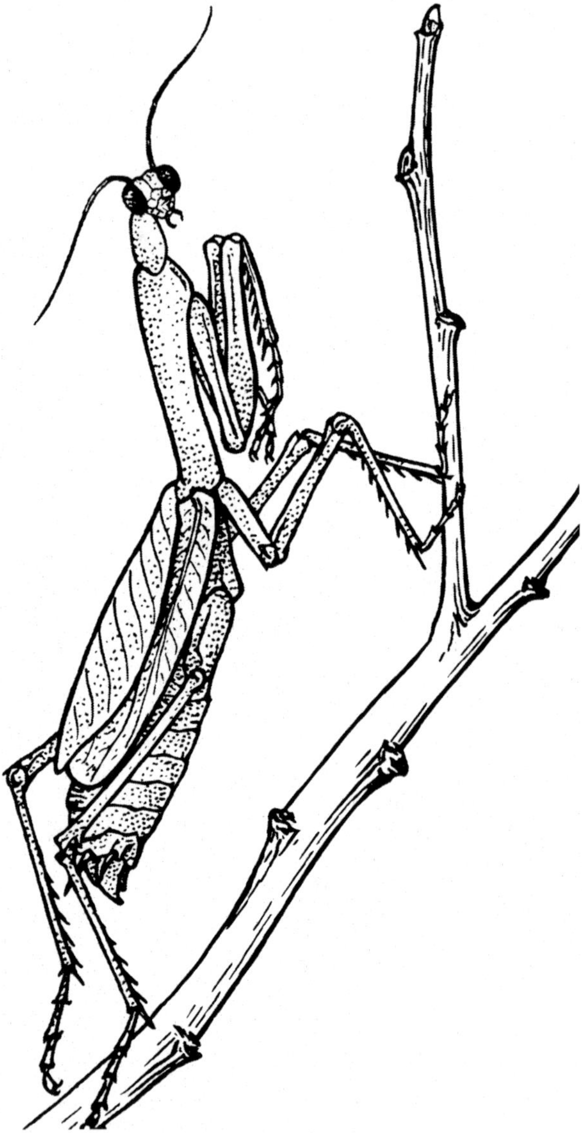

Praying Mantis

For dessert, the last entry in this Cafè leBug is another favorite bug of Sawyer Wright – the Praying Mantis. Mantises are an order of insects that contains 33 families with about 460 genera and over 2,400 species. Wow!

The largest family is the Mantidae or "mantids." Mantises are found all over the world in all climates except extreme cold. There are only 17 species found in North America. The 3 most common in the United States are European Mantis (Mantis *religiosa*), Chinese Mantis (Tenodera *sinensis*), and the Carolina Mantis (Stagmomantis *carolina*). The Carolina Mantis is the only native species of those 3.

Mantises have big, bulging, **compound** eyes on a triangular head supported by a long, flexible neck. They can turn their heads 180 degrees – a complete half circle. They have elongated bodies and may or may not have wings. All Mantodea have forelegs that are greatly enlarged and adapted for catching and gripping prey. which they usually eat head first. Their habit of remaining stationary in the upright posture with their forearms folded has led to their common name of "praying mantis."

We usually think of them as being green, but they actually come in all kinds of colors that help them blend in with the plant life in their habitats. Some species change color when they **molt** to match the plant life during different seasons. Some of them even have body shapes that help them look like branches and leaves which has caused them to mistaken for Stickbugs.

The praying mantis only lives about a year. They only mate once and this usually takes place in the fall. After the male and female have mated, the female will produce an **ootheca** – an egg mass that provides protection by encasing the eggs. There are between 100 and 200 eggs in the **ootheca**. The eggs develop over the winter months and the **larvae** hatch in the spring. The **larvae** will then undergo several **molts**, with the first happening shortly after hatch. Sadly, very few praying mantis **nymphs** survive the first **instar** stage due to the lack of food and cold temperatures. Also, the **larvae** will eat each other if there is not enough other food for them.

Mantises eat all different kinds of animals depending on their size. Insects, spiders, small lizards and frogs are all on their menu as well as small species of bird – like hummingbirds, flycatchers, warblers, and others. They only eat living creatures. As well as eating many forms of animal, they are also eaten by many forms of life – many are the same that they eat, like birds, lizards, frogs, and spiders. Female praying mantises often eat the males after mating. Between 13 and 28 percent of mating encounters end with sexual **cannibalism**, in which the female praying mantis bites off the head of the male and eats him. Studies have found that females who cannibalized their male partner produced significantly more eggs than those that did not, suggesting that their cannibalistic behavior may increase the chance of reproductive success. It sounds like it's a "dog eat dog world" for them!

What am I? Praying Mantis

What is my Body Type?

Hard [X] Segmented [X] How many Segments?

Soft [ ] Worm like [ ] 2 [ ] 3 [X] More ____

Do I have Legs? Yes [X] How many? 6 No [ ]

Do I have Wings? Yes [X] How many? 4 No [ ]

Are my Wings under ***Elytra***? Yes [ ] No [X]

Do I have Stripes? Yes [ ] No [X] How Many? ____

Do I have Spots? Yes [ ] No [X] How Many? ____

What Color am I? green My Spots or Stripes? ______

Kingdom Animalia

Phylum Arthropoda

Class Insecta

Order Mantodea

Family Mantidae

Genus Mantis

Species M. religiosa

Draw Me

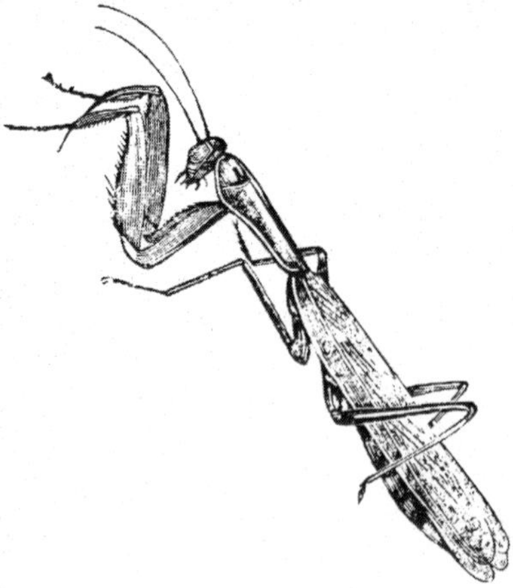

What am I? ____________

What is my Body Type?

Hard ☐ Segmented ☐ How many Segments?

Soft ☐ Worm like ☐ 2 ☐ 3 ☐ More ____

Do I have Legs? Yes ☐ How many? ______ No ☐

Do I have Wings? Yes ☐ How many? ______ No ☐

Are my Wings under ***Elytra***? Yes ☐ No ☐

Do I have Stripes? Yes ☐ No ☐ How Many? ____

Do I have Spots? Yes ☐ No ☐ How Many? ____

What Color am I? ______ My Spots or Stripes? ______

Kingdom ________

Phylum ________

Class ________

Order ________

Family ________

Genus ________

Species ________

Draw Me

What am I? ____________

What is my Body Type?

Hard ☐ Segmented ☐ How many Segments?

Soft ☐ Worm like ☐ 2 ☐ 3 ☐ More ___

Do I have Legs? Yes ☐ How many? ______ No ☐

Do I have Wings? Yes ☐ How many? ______ No ☐

Are my Wings under ***Elytra***? Yes ☐ No ☐

Do I have Stripes? Yes ☐ No ☐ How Many? ___

Do I have Spots? Yes ☐ No ☐ How Many? ___

What Color am I? ______ My Spots or Stripes? ______

Kingdom __________

Phylum __________

Class __________

Order __________

Family __________

Genus __________

Species __________

Draw Me

What am I? ________________

What is my Body Type?

Hard ☐ Segmented ☐ How many Segments?

Soft ☐ Worm like ☐ 2 ☐ 3 ☐ More ____

Do I have Legs? Yes ☐ How many? ______ No ☐

Do I have Wings? Yes ☐ How many? ______ No ☐

Are my Wings under ***Elytra***? Yes ☐ No ☐

Do I have Stripes? Yes ☐ No ☐ How Many? ____

Do I have Spots? Yes ☐ No ☐ How Many? ____

What Color am I? ______ My Spots or Stripes? ______

Kingdom __________

Phylum __________

Class __________

Order __________

Family __________

Genus __________

Species __________

Draw Me

What am I? ________________

What is my Body Type?

Hard ☐ Segmented ☐ How many Segments?

Soft ☐ Worm like ☐ 2 ☐ 3 ☐ More ____

Do I have Legs? Yes ☐ How many? ______ No ☐

Do I have Wings? Yes ☐ How many? ______ No ☐

Are my Wings under ***Elytra***? Yes ☐ No ☐

Do I have Stripes? Yes ☐ No ☐ How Many? ____

Do I have Spots? Yes ☐ No ☐ How Many? ____

What Color am I? ______ My Spots or Stripes? ______

Kingdom ________

Phylum ________

Class ________

Order ________

Family ________

Genus ________

Species ________

Draw Me

What am I? ____________

What is my Body Type?

Hard ☐ Segmented ☐ How many Segments?

Soft ☐ Worm like ☐ 2 ☐ 3 ☐ More ____

Do I have Legs? Yes ☐ How many? ______ No ☐

Do I have Wings? Yes ☐ How many? ______ No ☐

Are my Wings under ***Elytra***? Yes ☐ No ☐

Do I have Stripes? Yes ☐ No ☐ How Many? ____

Do I have Spots? Yes ☐ No ☐ How Many? ____

What Color am I? ______ My Spots or Stripes? ______

Kingdom ________

Phylum ________

Class ________

Order ________

Family ________

Genus ________

Species ________

Draw Me

What am I? ____________

What is my Body Type?

Hard ☐ Segmented ☐ How many Segments?

Soft ☐ Worm like ☐ 2 ☐ 3 ☐ More ___

Do I have Legs? Yes ☐ How many? ______ No ☐

Do I have Wings? Yes ☐ How many? ______ No ☐

Are my Wings under ***Elytra***? Yes ☐ No ☐

Do I have Stripes? Yes ☐ No ☐ How Many?___

Do I have Spots? Yes ☐ No ☐ How Many?___

What Color am I?______ My Spots or Stripes?______

Kingdom __________

Phylum __________

Class __________

Order __________

Family __________

Genus __________

Species __________

Draw Me

What am I? ____________

What is my Body Type?

Hard ☐ Segmented ☐ How many Segments?

Soft ☐ Worm like ☐ 2 ☐ 3 ☐ More ___

Do I have Legs? Yes ☐ How many? ______ No ☐

Do I have Wings? Yes ☐ How many? ______ No ☐

Are my Wings under ***Elytra***? Yes ☐ No ☐

Do I have Stripes? Yes ☐ No ☐ How Many? ____

Do I have Spots? Yes ☐ No ☐ How Many? ____

What Color am I? ______ My Spots or Stripes? ______

Kingdom __________

Phylum __________

Class __________

Order __________

Family __________

Genus __________

Species __________

Draw Me

What am I? ____________

What is my Body Type?

Hard ☐ Segmented ☐ How many Segments?

Soft ☐ Worm like ☐ 2 ☐ 3 ☐ More ____

Do I have Legs? Yes ☐ How many? ______ No ☐

Do I have Wings? Yes ☐ How many? ______ No ☐

Are my Wings under ***Elytra***? Yes ☐ No ☐

Do I have Stripes? Yes ☐ No ☐ How Many? ____

Do I have Spots? Yes ☐ No ☐ How Many? ____

What Color am I? ______ My Spots or Stripes? ______

Kingdom __________

Phylum __________

Class __________

Order __________

Family __________

Genus __________

Species __________

Draw Me

What am I? ____________

What is my Body Type?

Hard ☐ Segmented ☐ How many Segments?

Soft ☐ Worm like ☐ 2 ☐ 3 ☐ More ____

Do I have Legs? Yes ☐ How many? ______ No ☐

Do I have Wings? Yes ☐ How many? ______ No ☐

Are my Wings under ***Elytra***? Yes ☐ No ☐

Do I have Stripes? Yes ☐ No ☐ How Many? ____

Do I have Spots? Yes ☐ No ☐ How Many? ____

What Color am I? ______ My Spots or Stripes? ______

Kingdom ________

Phylum ________

Class ________

Order ________

Family ________

Genus ________

Species ________

Draw Me

What am I? ________________

What is my Body Type?

Hard ☐ Segmented ☐ How many Segments?

Soft ☐ Worm like ☐ 2 ☐ 3 ☐ More ___

Do I have Legs? Yes ☐ How many? ______ No ☐

Do I have Wings? Yes ☐ How many? ______ No ☐

Are my Wings under ***Elytra***? Yes ☐ No ☐

Do I have Stripes? Yes ☐ No ☐ How Many? ___

Do I have Spots? Yes ☐ No ☐ How Many? ___

What Color am I? ______ My Spots or Stripes? ______

Kingdom __________

Phylum __________

Class __________

Order __________

Family __________

Genus __________

Species __________

Draw Me

What am I? ____________

What is my Body Type?

Hard ☐ Segmented ☐ How many Segments?

Soft ☐ Worm like ☐ 2 ☐ 3 ☐ More ____

Do I have Legs? Yes ☐ How many? ______ No ☐

Do I have Wings? Yes ☐ How many? ______ No ☐

Are my Wings under ***Elytra***? Yes ☐ No ☐

Do I have Stripes? Yes ☐ No ☐ How Many? ____

Do I have Spots? Yes ☐ No ☐ How Many? ____

What Color am I? ______ My Spots or Stripes? ______

Kingdom ________

Phylum ________

Class ________

Order ________

Family ________

Genus ________

Species ________

Draw Me

What am I? ____________

What is my Body Type?

Hard ☐ Segmented ☐ How many Segments?

Soft ☐ Worm like ☐ 2 ☐ 3 ☐ More ___

Do I have Legs? Yes ☐ How many? ______ No ☐

Do I have Wings? Yes ☐ How many? ______ No ☐

Are my Wings under ***Elytra***? Yes ☐ No ☐

Do I have Stripes? Yes ☐ No ☐ How Many? ___

Do I have Spots? Yes ☐ No ☐ How Many? ___

What Color am I? ______ My Spots or Stripes? ______

Kingdom __________

Phylum __________

Class __________

Order __________

Family __________

Genus __________

Species __________

Draw Me

What am I? ________________

What is my Body Type?

Hard ☐ Segmented ☐ How many Segments?

Soft ☐ Worm like ☐ 2 ☐ 3 ☐ More ___

Do I have Legs? Yes ☐ How many? ______ No ☐

Do I have Wings? Yes ☐ How many? ______ No ☐

Are my Wings under ***Elytra***? Yes ☐ No ☐

Do I have Stripes? Yes ☐ No ☐ How Many? ____

Do I have Spots? Yes ☐ No ☐ How Many? ____

What Color am I? ______ My Spots or Stripes? ______

Kingdom ________

Phylum ________

Class ________

Order ________

Family ________

Genus ________

Species ________

Draw Me

What am I? ______________

What is my Body Type?

Hard ☐ Segmented ☐ How many Segments?

Soft ☐ Worm like ☐ 2 ☐ 3 ☐ More ___

Do I have Legs? Yes ☐ How many? ______ No ☐

Do I have Wings? Yes ☐ How many? ______ No ☐

Are my Wings under ***Elytra***? Yes ☐ No ☐

Do I have Stripes? Yes ☐ No ☐ How Many? ___

Do I have Spots? Yes ☐ No ☐ How Many? ___

What Color am I? ______ My Spots or Stripes? ______

Kingdom ________

Phylum ________

Class ________

Order ________

Family ________

Genus ________

Species ________

Draw Me

What am I? ______________

What is my Body Type?

Hard ☐ Segmented ☐ How many Segments?

Soft ☐ Worm like ☐ 2 ☐ 3 ☐ More ___

Do I have Legs? Yes ☐ How many? ______ No ☐

Do I have Wings? Yes ☐ How many? ______ No ☐

Are my Wings under ***Elytra***? Yes ☐ No ☐

Do I have Stripes? Yes ☐ No ☐ How Many? ___

Do I have Spots? Yes ☐ No ☐ How Many? ___

What Color am I? ______ My Spots or Stripes? ______

Kingdom ________

Phylum ________

Class ________

Order ________

Family ________

Genus ________

Species ________

Draw Me

Glossary

abdomen: the end part of the body of an arthropod.

antenna: a pair of long, thin sensory feelers on the heads of insects; singular - antennae.

cannibalism: the eating of any creature by another member of the same species.

carnivore: meat eater.

carrion: dead and rotting flesh.

cepholothorax: combined head and thorax into one body part.

cerci: a pair of pincers on the end of the abdomen; singular cercus.

compound: more than one thing joined together - like compound eyes are a clump of eyes.

convergent: coming together.

crustacea: mostly water-dwelling arthropods having an exoskeleton of chitin - a shell-like material made of a modified carbohydrate (polysaccharide).

diploid: the presence of two complete sets of chromosomes in an organism's cells, with each parent contributing one set of chromosomes to each pair.

elytra: the tough fore wings of beetles, elytra are not used in flight but are used to protect the more delicate hind wings.

exoskeleton: rigid or jointed casing (shell) that supports and protects the soft tissues of certain animals; like arthropods.

fore-wing: one of the front wings of a four-winged insect.

haploid: the presence of only one set of chromosomes in an organism's cells.

herbivore: plant eater.

instar: the larva or nymph existing between two successive molts.

mesothorax: the middle of the three segments of the ***thorax*** of an insect.

metathorax: the back of the three segments of the ***thorax*** of an insect.

misnomer: wrongly named.

molt: the process of shedding the exoskeleton of the previous life stage.

palps: (***pedipalps***) - a pair of sensory appendages at the mouthparts of arthropods.

parasite: an organism that lives on or in a host organism and gets its food from or at the expense of its host.

pedicel: the thin structure that connects the ***thorax*** and ***abdomen***.

proboscis: a long, thin tube that forms part of the mouth of some insects.

prothorax: the foremost of the three segments in the ***thorax*** of an insect, and bears the first pair of legs.

pterothorax: the ***mesothorax*** and ***metathorax*** in insects (fused in dragonflies), which carry the wings.

pupa: case or *chrysalis* in which insects pupate.

pupate: complete metamorphosis - change - of the insect body between the larval and adult stages.

omnivore: eats both plants and meat.

ootheca: protective egg case

ovipositor: a tubular structure attached to the ***abdomen*** of insects that is used for laying eggs.

salivary: glands that produce ***saliva*** - spit that moistens food to help digestion.

spinneret: a silk producing organ possessed by spiders and some insects.

tympanal organ: a hearing organ in insects consisting of a membrane (***tympanum***) stretched across a frame backed by an air sac.

thorax: the middle segment of the three main body parts of an insect, behind the ***head*** and before the ***abdomen***.

vegetative: plant matter.

vestigial: reduced or shrunken so that it can no longer function in its original capacity.

voracious: having a huge appetite; ravenous, insatiable.

Acknowledgements

I would like to thank my parents for bringing me into this wonderful diverse world.

I would like to thank my elementary teachers and librarian who introduced me to the life sciences.

I would like to thank the creators of shows like Wild Kingdom, and all the National Geographic specials that showed me the wonders of life on this planet.

I would like to thank my professors of Biology, Zoology, Entomology, and Botany who taught me more about the life on this planet and how it all works together.

I would like to thank Sawyer Wright for giving me inspiration.

I would like to thank my wonderful wife for putting up with me and helping me out in any way she can. Without her none of this would be possible for me.

REFERENCES

I did most of my research on the internet. Here are some of the sites I used:

https://animaldiversity.org/

https://en.wikipedia.org/wiki/Main_Page

https://www.insectidentification.org/

https://www.montana.edu/yellowstoneinsects/contents.html

ABOUT THE AUTHOR

Sam Humphrey

Sam was born in Moscow, Idaho and raised in the northwestern United States. He spent summers camping, hiking, and canoeing. He showed an interest in nature and life from a very young age. He was always looking at bugs, insects, arachnids, and worms. As an adult, he studied at several community colleges around Washington state and eventually earned a degree in Biology at Whitworth College in Spokane. He took classes in Zoology, Entomology, and other sciences. He was always looking at bugs. He spent a lot of time riding motorcycles which gave him a new appreciation for flying insects. He currently lives in Buffalo, Wyoming with his wonderful wife and two cats. He still looks at bugs and shares his interests in entomology and zoology with his grand nephews and nieces.

Made in United States
North Haven, CT
04 November 2026

21768990R00095